Industrial Numerical Analysis

PROFESSOR LESLIE FOX

Industrial Numerical Analysis

Edited by

SEAN McKEE

*Coordinator of the University Consortium for
Industrial Numerical Analysis*

and

CHARLES M. ELLIOTT

*Department of Mathematics, Imperial College,
University of London*

CLARENDON PRESS · OXFORD · 1986

Oxford University Press, Walton Street, Oxford OX2 6DP
Oxford New York Toronto
Delhi Bombay Calcutta Madras Karachi
Kuala Lumpur Singapore Hong Kong Tokyo
Nairobi Dar es Salaam Cape Town
Melbourne Auckland
and associated companies in
Beirut Berlin Ibadan Nicosia

Oxford is a trade mark of Oxford University Press

Published in the United States
by Oxford University Press, New York

British Library Cataloguing in Publication Data
Industrial numerical analysis.
1. Numerical analysis
I. Elliott, Charles M.   II. McKee, Sean
519.4'02462   QA297
ISBN 0 19 853190 7

Library of Congress Cataloging in Publication Data
Industrial numerical analysis.
Bibliography: p.
1. Numerical analysis.   2. Engineering – Mathematics.
I. Elliott, Charles M.   II. McKee, Sean.
TA335.I565   1985   620'.001511   85-21601
ISBN 0 19 853190 7

Printed in Great Britain by
St. Edmundsbury Press,
Bury St. Edmunds, Suffolk

Foreword

Over the last fifteen years there has been a marked change in the research activities of many applied mathematicians and numerical analysts. Instead of a concentration on more refined methods for treating established problems, there has been a rapid growth of interest in setting up links with industry and tackling a wide variety of completely new industrial problems.

Much of the stimulus for this activity has come from the Oxford Study Groups with Industry (OSGI) started in 1968 and the University Consortium for Industrial Numerical Analysis (UCINA) set up in 1979. This book is one of the results and half the problems treated here have arisen directly from their efforts. The variety of problem sources and of solution techniques is typical of this kind of research. It also provides a continual prompting to those engaged in more traditional research to see that their methods and understanding can encompass the new situations thrown up. This is most welcome.

The editors have performed an excellent service to the community in assembling such an interesting selection of material and I wish the volume a wide readership in both the academic and industrial communities.

Computing Laboratory
University of Oxford

Professor K.W. Morton

Contents

Contributors   ix

Introduction – *The Editors*   xi

1. The mathematical and numerical assessment of the flatness of engineering surface plates and tables *M.G. Cox*   1

2. A simple new design formula for chain links
    G.T. Anthony, T.A.E. Gorley, and J.G. Hayes   25

3. Optimal control problems in tidal power generation
    N.R.C. Birkett and N.K. Nichols   53

4. Low thrust satellite trajectory optimization
    L.C.W. Dixon, S.E. Hersom, and Z.A. Maany   90

5. A wire-upwinding problem *B. Benjamin and
    D. Handscomb*   108

6. Analysis of a wave power device *B.M. Count and
    C.M. Elliott*   124

7. Use of the linear functional strategy for assessing the status of plant canopies *R.S. Anderssen and
    D.R. Jackett*   143

8. The ageing of stainless steel *J. Norbury*   165

9. The development of mathematical models in welding and their numerical solution *J.G. Andrews and R.E. Craine*   184

10. A numerical model of the glass sheet and fibre updraw processes *C. Saxelby and J.M. Aitchison*   199

11. The motion of a towed array of hydrophones
    R. Cartwright and D.F. Mayers   216

12. A mathematical model of electromagnetic river gauging
    P. Jacoby and M. Baines   229

Contributors

J.M. Aitchison
Department of Mathematics, Royal Military College of Science,
Shrivenham, Swindon, Wiltshire

R.S. Anderssen
CSIRO, Division of Mathematics and Statistics, P.O. Box 1965,
Canberra ACT 2601, Australia

J.G. Andrews
Marchwood Engineering Laboratory, Central Electricity Generating
Board, Marchwood, Southampton, SO4 4ZB

G.T. Anthony
National Physical Laboratory, Division of Information Technology
and Computing, Teddington, Middlesex, TW11 0LW

M. Baines
Department of Mathematics, University of Reading, Whiteknights,
Reading, Berkshire, RG6 2AX

B. Benjamin
School of Mathematics, South Australian Institute of Technology,
P.O. Box 1, Ingle Farm, South Australia

N.R.C. Birkett
Department of Mathematics, University of Reading, Whiteknights,
Reading, Berkshire, RG6 2AX

R. Cartwright
Department of Materials Science and Engineering, Room 8-135,
Massachusetts Institute of Technology, Cambridge, Massachusetts
02139, U S A

B.M. Count
Marchwood Engineering Laboratory, Central Electricity Generating
Board, Marchwood, Southampton, SO4 4ZB

M.G. Cox
National Physical Laboratory, Division of Information Technology
and Computing, Teddington, Middlesex, TW11 0LW

R.E. Craine
Faculty of Mathematical Studies, University of Southampton,
Southampton, SO9 5NH

L.C.W. Dixon
Numerical Optimisation Centre, The Hatfield Polytechnic, P.O.
Box 109, College Lane, Hatfield, Hertfordshire, AL10 9AB

C.M. Elliott
Department of Mathematics, Imperial College, Queen's Gate,
London, SW7 2BZ

T.A.E. Gorley
National Physical Laboratory, Division of Information Technology
and Computing, Teddington, Middlesex, TW11 0LW

D. Handscomb
Computing Laboratory, University of Oxford, 8-11 Keble Road,
Oxford, OX1 3QD

J.G. Hayes
National Physical Laboratory, Division of Information Technology
and Computing, Teddington, Middlesex, TW11 0LW

S.E. Hersom
Numerical Optimisation Centre, The Hatfield Polytechnic, P.O.
Box 109, College Lane, Hatfield, Hertfordshire, AL10 9AB

D.R. Jackett
CSIRO, Division of Mathematics and Statistics, P.O. Box 1965,
Canberra ACT 2601, Australia

P. Jacoby
Computing Service, University of Leeds, Leeds

Z.A. Maany
Numerical Optimisation Centre, The Hatfield Polytechnic, P.O.
Box 109, College Lane, Hatfield, Hertfordshire, AL10 9AB

D.F. Mayers
Computing Laboratory, University of Oxford, 8-11 Keble Road,
Oxford, OX1 3QD

S. McKee
Computing Laboratory, University of Oxford, 8-11 Keble Road,
Oxford, OX1 3QD

K.W. Morton
Computing Laboratory, University of Oxford, 8-11 Keble Road,
Oxford, OX1 3QD

N.K. Nichols
Department of Mathematics, University of Reading, Whiteknights,
Reading, Berkshire, RG6 2AX

J. Norbury
The Mathematical Institute, University of Oxford, 24-29 St Giles,
Oxford, OX1 3LB

C. Saxelby
GCHQ, Cheltenham, Wiltshire

Introduction

The application of mathematics to industrial problems involves:

(a) The formulation of problems which are amenable to mathematical investigation.

(b) Mathematical modelling.

(c) The solution of the mathematical problem.

(d) The interpretation of the results.

These steps are not clearly separated; the solution of the mathematical problem may lead to a refinement in the mathematical model or a change in the nature of the questions to be tackled by mathematical methods. Diagram 1 shows the differing and interconnecting stages of mathematical problem solving.

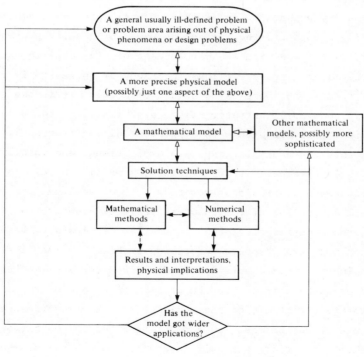

Diagram 1

The problem will often initially be a vague question. What should the thickness of a magnetic screen be in order to reduce the magnetic flux density to an acceptable amount? What is the best strategy for using turbines and sluices in the proposed Severn barrage in order to maximise the energy output? How should the process of cooling fish be controlled in order to ensure thorough freezing? Such problems are usually posed by a team of engineers and scientists, often with the aid of a mathematician. When a precise physical formulation of the problem has been obtained it must be approximated by a mathematical model. Often there will be several competing models of varying degrees of sophistication. Physical intuition and experience are invaluable to the modelling phase. The ideal is to study the simplest model which accurately describes the aspects of the problem in which one is interested. Occasionally numerical methods are unnecessary and analytical techniques will provide adequate understanding. However, the increasing power of readily available computers coupled with the development of high quality numerical algorithms and software allows the study of increasingly more sophisticated and complex mathematical models. It is the purpose of this book to illustrate, using case histories, the role of numerical analysis within this problem solving framework. In doing this the book also exposes computational and modelling techniques together with interesting, sometimes novel, physical and mathematical problems from a variety of applications.

The ordering of the chapters reflects a desire to group them around different topics in Numerical Analysis. The book begins with two chapters describing problems in metrology and approximation. In Chapter 1, Dr Cox discusses a scientific, computer-based approach to the problem of assessing the departure from perfect flatness of the working surface of engineering surface plates and tables. An account is given of the different types of data grids over which measurements are taken, and simple

mathematical models are built of two types of measuring instrument commonly used in the assessment. Systems of over-determined
linear equations are derived that relate gathered measurements
to heights which provide a discrete representation of the shape
of the working surface of the table. Details of the least
squares solution of these sparse equations are given. From results obtained it is shown how the use of an algorithm for linear
minimax approximations enables a table to be graded according to
national or international metrological standards.

Lifting gear is required to meet certain standards. In
particular chain links must comply to precise specifications. A
design formula had been in existence for some forty years and
the relevant committee of the British Standards Institution
decided that this should be updated. Drs Anthony, Gorley and
Hayes, in Chapter 2, discuss how this was achieved. Essentially
the problem is one of multivariate approximation with the requirement that the design formula be as simple and straightforward as
possible.

The next two chapters are concerned with the numerical
solution of optimal control problems. During the last twenty
years there has been considerable interest in developing alternative energy sources. In Chapter 3, Drs Birkett and Nichols
consider problems arising from tidal power generation by means
of a barrage across the River Severn and develop techniques for
determining the maximal average energy output using optimal control theory. The first model provides a simple test example in
which power is extracted from an oscillating system while the
second model simulates tidal power generation from flow across a
tidal barrage which contains both turbines and sluices. Numerical methods for computing solutions are derived and the predicted
optimal strategies are shown to be often in contradiction to
intuition.

The European Space Agency is clearly interested in satellite orbits which minimize the consumption of fuel. In Chapter 4

Drs Dixon, Hersom and Maany address this question and, while obtaining reasonably satisfactory results, highlight the fact that existing optimization routines are still inadequate.

The next four chapters are concerned with ordinary differential and integral equations. During the manufacture of wire it is stacked before, at a later time, being drawn off upwards and wound onto a drum. It is of importance to be able to do this quickly without snagging. Drs Benjamin and Handscomb discuss this problem in Chapter 5. Mathematically it is easy to see that the problem is a two point boundary value problem with one of the unknowns being the length of the wire; however, it is not clear which choice of boundary conditions will make the problem well-posed. An extensive discussion is presented detailing the avenues taken in attempting to resolve these difficulties and the results so far obtained.

Energy can be extracted from the sea using the motion of structures excited by waves. In Chapter 6, Drs Count and Elliott derive a novel finite difference scheme for solving an integro-differential equation (strictly an integro-differential inclusion) which arises from modelling the motion of a particular wave energy device, namely Salter's duck. The point of interest here for the numerical analyst is that for some wave trains the duck will remain stationary for unknown intervals of time. This can induce 'numerical chatter' in standard ODE solvers.

Chapter 7 describes the use of mathematical and numerical analysis in forest management and environmental modelling. It is of interest to discover the orientation of leaves of trees in a forest. A mathematical model of this problem leads to the study of an ill-posed linear Fredholm integral equation. Drs Anderssen and Jackett show how the use of linear functionals leads to the extraction of useful information from an ill-posed problem which has errors in its data.

The nuclear industry has a particular interest in the

ageing of steel. Stainless steel contains 15-20% of chrome and
0.1% of carbon and it is known that, at operating temperatures,
the chrome and the carbon precipitate out with the effect that
the steel is less able to withstand corrosive attack. A mathematical model consists of two, nonlinearly coupled, parabolic
equations with very different time constants. There were computational difficulties associated with the direct solution of the
parabolic equations because of the inherent stiffness of the
problem. In Chapter 8, Dr Norbury shows how to reduce the problem to a single nonlinear singular Volterra integral equation of
the second kind and, after a novel transformation, goes on to
solve it using a form of product integration. The results were
compared with experiments carried out at Berkeley Nuclear Research
Laboratory and good agreement was obtained; thus veriṯying the
validity of the mathematical model and the reduced equations.

The final four chapters are concerned with the numerical
solution of partial differential equations. Chapter 9 is concerned with welding. Drs Andrews and Craine derive a mathematical model of the temperature and fluid flow which leads to a free
boundary problem. Body fitted coordinates are used to map the
weld pool into a hemisphere with a fixed radius. A solution is
then sought through a finite difference technique. It is known
from experiments that the depth of a weld pool is greater than
its width; this is predicted by the model.

Before the invention of the float glass technique, glass
was manufactured by an updraw process. This is still used today
to make glass rods and is a feasible process for the manufacture
of glass fibres. In Chapter 10, Drs Saxelby and Aitchison have
considered a free boundary problem for the Stoke's equations
both in planar and cylindrical geometries. Using a finite
element Galerkin approach the authors have solved a sequence of
problems in fixed domains by iterating on the position of the
free surface. Although the model appears to give good agreement

in practice, the glass industry would be interested in a more sophisticated model which took account of the effects of surface tension and the temperature dependent viscosity of the glass.

The Ministry of Defence has long been interested in the detection of sound waves under water. One way to achieve this is by towing a hydrophone from a ship's stern. However, for a ship to maintain a steady course it must continually be correcting its steering to counter the wind and sea currents. This will induce an oscillatory motion in one end of the towed array; the question then arises as to the stability of the far end. Drs Cartwright and Mayers, in Chapter 11, illustrate that this can be modelled by an interesting nonlinear hyperbolic equation which can be solved by a finite difference method and the method of character- istics. The numerical experiments show that the motion is stable provided a drogue is fitted to the tail end of the hydrophone.

The continuous measurement of the volumetric flow of water in a river is important, not least to predict the possibility of flooding; sluices can then be opened to remove excess water to canals or to arable land far enough away from urban development. Electromagnetic water gauging makes use of the fact that a magne- tic field will interact with a moving conductor of electricity, that is, water. This induces a potential difference which can be measured. In Chapter 12, Drs Jacoby and Baines consider a mathematical model, the solution of which can be used to pro- vide a calibration for the gauge. They discuss the approximation of the three dimension Laplace's equation using finite element techniques.

Around half the problems of this book have come either from the Oxford Study Groups with Industry (OSGI) or the University Consortium for Industrial Numerical Analysis (UCINA). The Oxford Study Groups with Industry were started by Dr A.B. Tayler and Professor L. Fox and are now run by Dr J.R. Ockendon. Their principle concern is with the mathematical modelling of industrial

problems. The problems of chapters 6 and 11 came directly
through OSGI's industrial contacts. The University Consortium
for Industrial Numerical Analysis came about initially from a
Science and Engineering Research Council grant proposal of
Professor L. Fox and is more concerned with the numerical analysis of industrial problems. The coordinator is the second
editor of this book and, like Dr Ockendon, is a member of the
University of Oxford. However, UCINA is a wider organisation
embracing, not just Oxford, but also the Universities of Bath,
Reading and Swansea, Brunel University, Imperial College of
Science and Technology and the Division of Information Technology
and Computing of the National Physical Laboratory. The problems
of Chapters 3, 5, 8 and 10 came from UCINA while those from
Chapters 1 and 2 come directly from the National Physical Laboratory. Chapter 4 describes work done at the Hatfield Optimization
Centre while Chapters 7, 9 and 12 arose out of the personal contacts of the editors and the authors.

We would like to express our gratitude for the care taken
by our typist, Mrs Ina Godwin, and our illustrator, Mrs Marion
Stockton.

Finally we take great pleasure in dedicating this book to
our old friend and colleague, Professor Leslie Fox, for his
distinguished record in encouraging academic/industrial collaboration.

1

The mathematical and numerical assessment of the flatness of engineering surface plates and tables

M. G. COX

1. INTRODUCTION

The nominally flat surface plates and tables used widely throughout industry for marking out, inspection and quality assurance purposes are calibrated at manufacture and periodically thereafter. Calibration assesses their departure from perfect flatness and enables the tables to be graded as regards their suitability for various categories of work. Many manufacturing contracts, vital to industry, depend upon the demonstrable quality of the equipment used. Flatness is a particularly important consideration because surface plates and tables are frequently used as a datum from which many measurements are made of engineering components mounted upon them.

A method for flatness assessment that is in widespread use is outlined in a British Standards Specification (BSI [1972]). The method, originally designed for hand calculation, is increasingly felt to be inadequate for current industrial purposes. Instrument makers have provided computer implementations, but have failed to take advantage of the power of modern computers to undertake large calculations rapidly, in that the algorithms they employ apparently do little more than mimic the hand calculation method.

This chapter describes an approach to the problem that

1

provides a scientific, computer-based evaluation. Nevertheless, computation time for the method is insignificant: on a popular minicomputer a full table assessment from several hundred readings can be made in a few seconds.

The calibration process consists initially of taking a representative set of readings in a systematic way in the working surface of the table using an appropriate measuring instrument. The readings are then submitted to hand or computer analysis to determine the overall 'shape' of the table, as well as the required measure of its departure from perfect flatness. From this measure a grading for the table can be determined in accordance with national or international metrological standards.

In addition to the British Standards Specification BS 817 (BSI [1972]) for surface plates, there are in existence other standards (German, American, International), but for our purposes their content can be regarded as differing only in points of detail from that of BS 817.

The history of flatness assessment, insofar as it is relevant to our discussion, is briefly as follows. The 1972 version of BS 817 was the third released, superseding versions dated 1938 and 1957. The hand-calculation assessment procedure described there is based upon taking observations over a 'Union Jack' grid of measurement lines laid down in the working surface of the table. (A new version of BS 817, dated 1983, has recently appeared. Fewer calculation details are given than in the 1972 edition, the reader being referred to the technical literature.) Deficiences in the procedure were recognized, and subsequent work at the National Physical Laboratory (Birch & Cox [1973]) led to an improved procedure which (a) utilized a generalization of the Union Jack grid, and (b) could be implemented on a digital computer using reliable numerical methods. The generalization consisted of a uniform rectangular grid with two additional, diagonal lines or generators. Further NPL work (Birch & Harrison [1974]) resulted in the development of methods for triangular, or hexagonal, grids, which in many circumstances have advantages over rectangular grids. After obtaining extensive experience in the mid-1970s with the use of these methods, work was started in 1977 on FLATPAK, a definitive Fortran package based upon them. The first version of FLATPAK was completed in 1979, and made available (Compeda [1980]) the following year. The package has now been used by a number of establishments covering British industry, academic institutions and government departments. This paper discusses the general approach we have adopted for flatness assessment, and describes the mathematical basis of FLATPAK and the numerical methods it employs.

The chapter is organized as follows. Section 2 describes the taking of observations, emphasizing that the use of different types of measuring instrument gives rise to different types of dimensional information. The subject of data grids or arrays — the structure of the set of measurement points — is addressed in Section 3. In Section 4, simple mathematical models are constructed which relate the gathered dimensional information to the required surface heights. The measurement procedures and, through the form of data grid and the mathematical model, the overdetermined systems of equations from which the surface heights are to be obtained, are considered in Section 5. In Section 6 the least squares solution of those equations is discussed. It is shown how account can be taken of the rank deficiency present in the associated observation matrix, and of the zero-nonzero structure of this matrix. Statistical considerations are the subject of Section 7, where the detection of erroneous measurements is discussed. In Section 8, a minimax approximation technique is used to determine the pair of parallel planes of minimal separations that contain between them the computed surface heights, from which a grading for the table is determined. Finally, Section 9 outlines the software implementation of the method.

2   TAKING OBSERVATIONS

    The first stage in flatness assessment is to lay down a
suitable measurement array or grid in the working surface of the
table. Observations are taken in a systematic way throughout the
grid using an appropriate measuring instrument. Instruments are
available for measuring (a) relative heights, i.e. height values
with respect to a certain datum, and (b) relative gradients, i.e.
gradient values, again with respect to a certain datum.

    Typical of a height-measuring instrument is the oblique
incidence interferometer, which provides a height estimate relative to the 'base line' of the instrument. An example of a
gradient-measuring instrument is the electronic precision level,
which yields the gradient of the chord joining two points (at
which are placed the 'feet' of the instrument), relative to the
setting of the controls of the instrument. The electronic level
does not permit such an accurate assessment as that provided by
the interferometer, but except possibly for the calibration of
the highest-precision inspection tables it is perfectly adequate.

3.   DATA GRIDS

    In order to obtain a distribution of readings over the
working surface, a grid or array of straight line generators is
first laid down. Measurements are then taken in a systematic
manner throughout the grid.

    We describe first the rectangular array, which is a
natural generalization of the traditional 'Union Jack' grid, a
measurement procedure for which is given in BS 817. The Union
Jack grid, discussed next, is used in some commercially available computer-aided flatness measurement systems. Finally we
describe triangular, or hexagonal, grids. Relative merits of
these grids are indicated.

    The $n \times n$ rectangular grid consists of three sets of
generators. The first set contains n (an odd number) uniformly
spaced lines parallel to the x-axis, the second n such lines

parallel to the y-axis, and the third two diagonal generators, joining opposite corners of the bounding rectangle so formed. Uniformly spaced points or 'stations' are placed along each grid line, the first and last stations being on the boundaries of the grid. The number of stations along a grid line is chosen to be the same for each line in a set, and is made one greater than an integral multiple of $n-1$, thus ensuring that there are stations at all vertices.

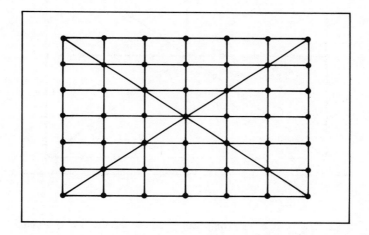

Figure 1   A generalized 'Union Jack', or rectangular, measurement grid with seven 'horizontal' and seven 'vertical' generators.

Figure 1 depicts a rectangular grid for which $n=7$. In this case the same integral multiple, viz. unity, is taken for each set of generators, and thus stations occur at and only at vertices. The station spacing is different for each set of generators.

By taking a rectangular grid with $n = 3$ we obtain the
Union Jack grid. In Figure 2 the stations are positioned by
taking appropriate integral multiples of $n - 1$ such that the
station spacing along gridlines is the same for all three sets.

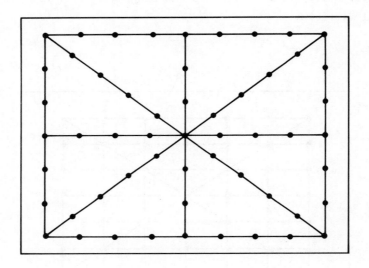

Figure 2   A 'Union Jack' measurement grid with
the same node spacing along all generators.

Unless the relative lengths of the generators in each of the
three sets can be expressed as the ratio of three integers
(e.g., 3:4:5, as in the figure), the spacing will not be the
same for the whole grid. This point is relevant in respect of
the practical aspects of data collection, as will be discussed
later.

The third type of array we discuss is the triangular, or hexagonal, array. The array is formed by three sets of uniformly spaced parallel generators. The spacing is the same for all three sets. The first set of generators we term the A-generators:

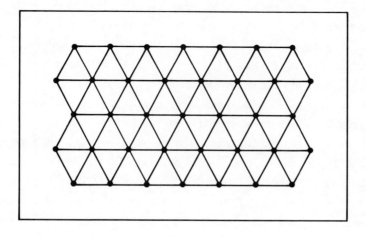

Figure 3 A triangular, or hexagonal, measurement grid with five rows, seven vertices on each odd-numbered row and eight vertices on each even-numbered row.

there is an odd number of these and in Figure 3 they run horizontally across the page. The second set, the B-generators, is at an angle of 120° to the first set (measured anti-clockwise). The third set, the C-generators, is at an angle of 60° to the first set, and located such that they pass through the vertices

formed by the A- and B-generators. The right-hand boundary of
the array is defined as follows. Starting at a vertex on the
top A-generator, trace a path between neighbouring vertices
by moving alternately along B- and C-generators, selecting a
B-generator first, until the bottom A-generator is reached. The
left-hand boundary is defined analogously, except that a C-
generator is selected first.

The resulting '$q \times r$' array is characterized by three parameters: q, the number of vertices in the first row (or in any
odd-numbered row), r, the number of rows (an odd number), and
h, the side of the basic equilateral triangle formed by the grid
lines. A station is placed at each vertex of the array. Figure
3 shows a triangular array with $q = 7$ and $r = 5$.

We conclude this section with a brief comparison of the
data grids. Apart from simplicity the Union Jack has little to
commend it. With the exception of special cases, such as that
depicted in Figure 2, the node spacing is different for the
'horizontal', 'vertical' and 'diagonal' grid lines. As a consequence, the gathering of data using a gradient-measuring instrument will necessitate remounting the instrument on bases of
different lengths during the measurement process. Moreover,
regardless of the node spacing along grid lines, the coverage of
the working surface of the table is highly nonuniform: there are
eight triangular regions (with sizes independent of the number
of nodes), each of which has an area equal to one eighth of the
embracing rectangle and whose interior is 'unsampled'. The use
of the generalized Union Jack, or rectangular, grid overcomes
the second objection, but not the first. For approximately the
same number of measurements, the triangular array not only provides a somewhat superior coverage in that it gives a uniform
density of points, but also, because of the constant distance
between neighbouring vertices, requires no change in the instrument base throughout the data acquisition stage when a gradient-
measuring instrument is used.

There are also statistical advantages in the use of the
triangular grid. As is shown in the following Section, for
'large' grids the ratio of the number of observations to the
number of surface heights approaches two for the rectangular
grid and three for the triangular grid. The consequent greater
redundancy in the latter case implies increased statistical confidence in the computed height estimates. A disadvantage of the
triangular grid is the additional time taken to mark out the
array prior to taking the measurements. We consider subsequently
both the rectangular and triangular grids.

4.   MODELLING THE PHYSICAL SITUATION

Let x, y, z denote a system of Cartesian coordinates with
$0x$ parallel to a longer side of the table, assumed rectangular,
$0y$ parallel to a shorter side, and $0z$ orthogonal to these.

Consider first the use of a height-measuring instrument
to gather readings along a generator identified by the index k,
say. Let the base line of the instrument, relative to which
measurements are made, be $z = a(k) + b(k)\xi(k)$, say, where $a(k)$
and $b(k)$ are regarded as (unknown) constants for the generator,
and $\xi(k)$ is the direction of alignment of the instrument in the
x,y plane. Let $d(k,P)$ be the height measurement at a point P
on generator k in the working surface of the table, $z(P)$ the
(unknown) height at P, and $e(k,P)$ the corresponding observation
error. Then the mathematical model or 'observation equation' for
the measurement is

$$d(k,P) = z(P) - a(k) - b(k)\xi(k,P) + e(k,P),$$

where $\xi(k,P)$ denotes the value of $\xi(k)$ at P.

Now consider the use of a gradient-measuring instrument
to gather readings along the generator. For a particular measurement the instrument is positioned such that its feet are at the
points P and Q, say, on the generator. Let the measurement
taken be denoted by $d(k,P,Q)$ and its error by $e(k,P,Q)$. Then
if h is the instrument baselength, i.e., the Euclidean distance

between P and Q, the 'measured' height difference is $hd(k,P,Q)$. The mathematical model or observation equation corresponding to the measurement is

$$d(k,P,Q) = (z(Q) - z(P))/h + e(k,P.Q). \qquad (4-1)$$

This observation equation is 'correct' only if the precision level is calibrated to give zero reading in the gravity plane. Since it is often inconvenient so to calibrate the instrument, a more general observation equation is used in which a calibration, or instrument, constant is added to the right-hand side of (4-1).

5.   MEASUREMENT PROCEDURES

In the case of the rectangular $n \times n$ grid (typified by Figure 1 in which $n = 7$) we take measurements in turn along the 'horizontal' generators, followed by those along the 'vertical', and then the diagonal generators. Suppose the vertices of the array are numbered sequentially row by row, left to right along each row. If an interferometer is used as the measuring instrument, the following general set of observation equations is obtained (in which the a's and b's denote 'instrument constants' into which the node spacing has been multiplicatively absorbed, and $t = \frac{1}{2}(n+1)$). (For simplicity a sequential indexing system for the d's and e's is used here and subsequently.)

'Horizontal' generators:
$$d_l = z_l - a_i - b_i(j - t) + e_l,$$
with
$$l = (i-1)n + j, \quad j = 1,\ldots,n; \quad i = 1,\ldots,n.$$

'Vertical' generators:
$$d_{n^2 + l} = z_{(i-1)n+j} - a_{n+j} - b_{n+j}(i - t) + e_{n^2 + l},$$
with
$$l = (j - 1)n + i, \quad i = 1,\ldots,n; \quad j = 1,\ldots,n.$$

Diagonal generators
$$\left. \begin{array}{l} d_{2n^2 + i} = z_{(n+1)i-n} - a_{2n+1} - b_{2n+1}(i-t) + e_{2n^2 + i} \\[2mm] d_{2n^2+n+i} = z_{(n-1)i+1} - a_{2n+2} - b_{2n+2}(i-t) + e_{2n^2 + n+i} \end{array} \right\} i = 1,\ldots,n.$$

The resulting set of $m = 2n(n+1)$ observation equations contains $p = (n+2)^2$ unknowns (n^2 height values and $4n+4$ 'instrument constants'). For large values of n the ratio m/p approaches two. Each equation contains just three unknowns — one surface height and two 'instrument constants'.

If an electronic level is used to take the measurements over the rectangular array, the following sequence of observation equations is obtained, in which c denotes a 'generic' parameter representing any one of the constants introduced by instrument resettings.

'Horizontal' generators:

$$d_{l-i+1} = (z_{l+1} - z_l)/h + c + e_{l-i+1},$$

with

$$l = n(i-1) + j, \quad j = 1, \ldots, n-1, \quad i = 1, \ldots, n.$$

'Vertical' generators:

$$d_k = (z_{l+n} - z_l)/h + c + e_k,$$

with l as above,

$$k = (n-1)(n-1+j) + i, \quad i = 1, \ldots, n-1, \quad j = 1, \ldots, n.$$

Diagonal generators

$$d_k = (z_{l+1} - z_{l-n})/h + c + e_k,$$

with

$$l = (n+1)i, \quad k = 2n(n-1) + i, \quad i = 1, \ldots, n-1.$$

$$d_k = (z_{l+n} - z_{l+1})/h + c + e_k,$$

with

$$l = (n-1)i, \quad k = (2n+1)(n-1) + i, \quad i = 1, \ldots, n-1.$$

The number of equations is $m = 2(n^2 - 1)$ and that of unknowns $p = n^2 + s$, where s is the number of constants introduced by instrument resettings (including one assumed initially). If the number of resettings is limited, for large problems the ratio m/p again approaches two. Each equation contains three unknowns — two surface heights and one 'instrument constant'.

For triangular arrays the corresponding systems are defined by a similar but more complicated recipe. In FLATPAK the equations are formed by subdividing them into seven groups.

The first group relates to A-generators, the next three groups to B-generators and the last three to C-generators. The B-generator equations are so grouped to cover generators spanning the array from (a) left to bottom, (b) top to bottom, and (c) top to right (and similarly for the C-generators). Calls to a simple subroutine are made in each of these seven cases to determine the variables appearing in the respective equations.

In the case of height measuring instruments there are $m = 3qr + \frac{3}{2}(r-1)$ equations and $p = qr + 4(q+r) + \frac{1}{2}(r-5)$ unknowns, whereas for the gradient measuring instrument $m = (r-1)(3q-\frac{1}{2}) + q - 1$ and $p = qr + \frac{1}{2}(r-1) + s$. In both cases there are again just three unknowns in each equation, but now for 'large' problems the ratio m/p approaches three.

6.   DETERMINATION OF THE SURFACE HEIGHTS

In terms of an obvious matrix notation, any of the above systems of observation equations can be expressed as

$$d = Az + e$$

where A is a known $m \times p$ matrix, d a known m-vector of 'observations', and z a p-vector of heights and instrument constants. We wish to determine the components of z such that $e^T e$ is a minimum. For numerical stability we employ orthogonalization techniques (cf. Golub [1965]), and therefore form an orthogonal matrix Q of order m and an upper triangular matrix R of order p such that

$$A^T = (R^T \quad 0)\, Q\,.$$

We let

$$d^T = (\theta^T \quad \theta'^T)\, Q\,,$$

where θ contains p elements and θ' has $m-p$ elements. Then (Golub [1965]) z is the solution to the triangular system

$$Rz = \theta \qquad\qquad (6-1)$$

and the minimum value of $e^T e$ is $\theta'^T \theta'$.

The solution to (6-1) is not unique, however. The reason is that since the Cartesian axis system imposed upon the problem

is quite arbitrary the solution can be determined only with respect to an arbitrary 'datum' or 'reference' plane. Thus because
there are three degrees of freedom associated with this plane,
the matrix A, and hence R, has rank $p-3$.

    Particular solutions to rank-deficient least squares
problems can be determined using for example the methods described by Peters & Wilkinson [1970] for the minimum norm solution
(that which minimizes $z^T z$ and which *is* unique). The use of
such methods is inappropriate for our problem, for two reasons.
First, they require, for numerical stability, column interchanges
during the orthogonal reduction of A to triangular form. Interchanges are undesirable because their use loses the advantages
of the structure (the zero-nonzero pattern) of A. Second, it is
perfectly satisfactory, in our problem, to determine *any* particular solution, because any other can then readily be obtained
by a planar shift of the resulting array of computed surface
heights. FLATPAK provides solutions 'corrected' to three reference planes commonly used in practice: (i) the 3-zeros plane (the
plane passing through three points in the array of computed surface heights),(ii) the least squares mean plane (that which has
smallest departure, in the sense of least squares, from the computed surface heights), (iii) the minimal deviation plane (that
which minimizes the maximum departure from the computed surface
heights).

    We thus seek an *arbitrary* solution. We may assign zero
(or any numerical constant) to any three of the surface height
values that correspond to noncollinear points in the (x,y)
plane, thus defining as an initial reference the plane taking
the assigned values at these points), and solve, uniquely, for
the remaining unknowns. Equivalently, and more straightforwardly
in practice, we may regard each such assignment as a *resolving
equation* or *constant* (Gentleman [1974a]), which is simply appen-
ded either to the system of observation equations or to the
equations (6-1) once formed. The resulting least squares problem

is of full rank, the additional equations each essentially providing a rank one updating of the system (6-1).

Each matrix A has special structure, account of which must be taken in order to yield an algorithm that is efficient in both space and time. If the first and last nonzero elements in row i of a rectangular matrix B are in columns p_i and q_i, respectively, B is termed a *band* matrix with bandwidth $b = \max_i(q_i - p_i + 1)$. If C is a rectangular matrix of the same dimensions as B, with nonzero elements only in its rightmost c columns, the matrix $B + C$ is termed a *bordered*, or *augmented band*, matrix with bandwidth b and columnwidth c. If $p_{i+1} \geqslant p_i$ for all i, the matrix is said to be in *standard form*. It is of course always possible to permute the rows of a band or bordered matrix to bring it into standard form.

If their columns are ordered such that the leftmost ones correspond to the surface heights and the remainder to the instrument constants, the matrices A in our problem are bordered matrices with bandwidths and columnwidths as specified in Table 1. (In exceptional circumstances A will be a band matrix. Such a case arises if a gradient measuring instrument is used, the instrument being initially calibrated to give zero reading in the gravity plane, and no subsequent resettings made.)

The application of orthogonal transformations to reduce such matrices A to triangular form R yields bordered triangular matrices with the same bandwidth and columnwidth as the original matrices (Cox [1981]). Careful use of these transformations avoids arithmetic operations on the (zero) elements outside the band and column part of these matrices. Further, initial row permutation of A to standard form yields distinct processing time advantages, the structure of R being unaffected. As derived in Cox [1981] the computing time for the reduction of a matrix in standard form is proportional to $m(b+c)^2$ whereas the expected computed time for one in nonstandard form is proportional to $mp(b+c)$. Figures $4-7$ give examples of A and the corresponding R.

Table 1 Matrix dimensions, bandwidths and columnwidths

Measuring instrument	Type of grid	Grid dimensions(1)	Matrix dimensions		Band-width	Column-width
			Rows	Columns(2)		
Height	Rectangular	$n \times n$	$2n(n+1)$	$(n+2)^2$	1	$4(n+1)$
Gradient	Rectangular	$n \times n$	$2(n^2-1)$	n^2+s	$n+1$	s
Height	Triangular	$q \times r$	$3qr + \frac{3}{2}(r-1)$	$qr+4(q+r)+\frac{1}{2}(r-5)$	1	$4(q+r)-2$
Gradient	Triangular	$q \times r$	$(r-1)(3q-\frac{1}{2})+q-1$	$qr+\frac{1}{2}(r-1)+s$	$q+2$	s

(1) Number of vertices in first row times number of rows

(2) s denotes the number of resettings of the instrument

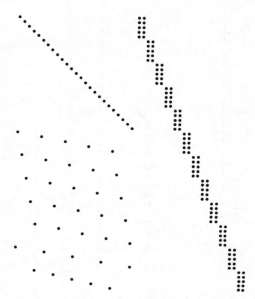

Figure 4   The structure of the observation matrix A (band-width 1 and columnwidth 24) for data gathered with a height-measuring instrument over a 5×5 rectangular array.

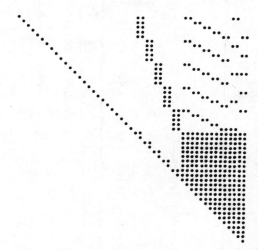

Figure 5   The structure of the triangular matrix R to which the observation matrix A of Figure 4 is reduced by ortho-gonal transformations.

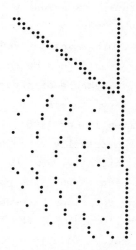

Figure 6 The structure of the observation matrix A (band-width 6 and columnwidth 3) for data gathered with a gradient measuring instrument over a 4×5 triangular array (with one instrument resetting for each of the A-, B-, and C-generators).

Figure 7 The structure of the triangular matrix R to which the observation matrix A of Figure 6 is reduced by ortho-gonal transformations.

For cases as small as these there would not be a great
deal to gain by regarding the observation and triangular matrices
as bordered. However, the figures do typify the structure of
matrices arising in flatness assessment problems. In larger flatness problems A may have 200 to 400 rows and 100 to 300 columns.
The processing time and storage space saved by taking advantage
of the bordered structure would then be considerable and, for
small computers, *essential* because of memory limitations, unless
use can be made of auxiliary storage.

The algorithms given in Cox [1981] for bordered least
squares problems, work sequentially through the rows of A, using
Givens plane rotations. Initially, both R and θ are set to zero.
Once each row of A has been formed, its nonzero elements are
annihilated by performing rotations between it and the rows of R.
The same rotations are performed upon the corresponding d-value
and the vector θ Each set of rotations to annihilate a row of
A essentially updates R and θ to give a triangular system, a
solution to which is a least squares solution to the set of observations processed so far. A valuable feature of this approach is
that A itself does not have to be stored. The method requires
storage only for the band and column parts of R, the vector θ
and the *current* row of A.

The matrices A evidently possess *fine* structure, i.e.
zero elements within their band and column parts. To a lesser
extent, the R-matrices also have fine structure, particularly in
the case of height-measuring instrument data, as indicated in the
figures. It is possible to exploit this fine structure in the
devising of flatness algorithms. Indeed, the structure of R can
be pre-determined (i.e. before any floating point computations
are carried out), using, for example, symbolic Givens rotations
(cf. Manneback [1983]), and calculations then performed only upon
elements of R which will ultimately be nonzero. Alternatively,
a general sparse matrix code, such as described by Duff & Reid
[1976] could be employed. The adoption of either of these

approaches should lead to a reduction in both real array storage
requirements and the total number of floating point operations
performed. Offsetting this reduction are the greater complexity
of the code and the integer arrays necessary to represent the
matrix structure. It is hoped that subsequent work will quantify
the possible savings.

7.   STATISTICAL CONSIDERATIONS

Once the least squares solution has been computed, the
residuals e_i of the model can be determined. Under the assumption that the gathered measurements have errors that are distributed normally with zero mean, the normalized residuals e_i/σ,
where σ is the standard deviation of the measurement errors,
are distributed as Student's t. Thus any unreasonably large
residuals can automatically be identified; the corresponding
observations may then be checked for gross errors. (An estimate
of σ is the root mean square residual $\|\theta'\|_2/(m-p)^{\frac{1}{2}}$.) This
simple device, although not foolproof, has proved to be valuable
in locating operator errors and equipment malfunctions.

If errors are suspected the corresponding measurements may
be re-taken, and modified data entered for analysis. Note that
for interferometer data, because the measurements depend upon
the position of the (unknown) baseline of the instrument, a single
erroneous observation necessitates the re-measurement of some or
all relative height values along the corresponding generator. In
the case of precision-level data, a number of the measurements
previously made for the same setting of the controls of the
instrument should be re-taken.

If the number of new observations is not too large, there
may be advantages in carrying out an updating of R by the new
observations and a downdating by the old ones (see Gill *et al.*
[1974] for methods). Such processes are equivalent to operating with matrices in nonstandard form, and are therefore not
particularly efficient. However, since, as stated, a typical

flatness computation takes only a few seconds on a minicomputer, resubmission of the complete set of appropriately modified data is therefore entirely reasonable. In our experience the time taken for input, output, production of graphs, etc., tends to swamp that for carefully designed mathematical software (cf. Gentleman [1974b]).

8. GRADING THE TABLE

The grading of the table, according to BS 817, is determined from the separation of two parallel planes just containing the working surface of the table. Such a measure is not directly implementable, however, because it depends upon the working surface itself — a continuum. We could, therefore, interpret this definition as relating to a finite set of *representative* points in the working surface, for which the parallel planes of minimal separation can be computed. Even this interpretation, which is the one used in FLATPAK, is open to criticism because we have at our disposal only best (least squares) estimates of such a set of points. The main point being emphasized here is that an algorithm and associated software can provide only an *interpretation* of the Standard. It is of crucial importance to the user of an algorithm to have confidence in the product. By stating clearly the limitations of an algorithm, such as those above, its designer is likely to promote such confidence.

At each vertex of the data array an estimated height z_i, say, is available. It is required to determine the plane

$$z = \alpha_1 + \alpha_2 x + \alpha_3 y$$

such that the maximum distance from it to the point z_i is minimized with respect to the plane parameters $\boldsymbol{\alpha} = (\alpha_1, \alpha_2, \alpha_3)^T$. This plane is termed the minimal deviation or minimax plane. The separating planes are then the planes parallel to the minimax plane which pass through the points of greatest positive and negative deviation.

For minimal separation, distance must be measured normally

to the plane. Thus the minimax plane is given by the solution
to the problem

$$\min_{\alpha} \max_{i} (1 + \alpha_2^2 + \alpha_3^2)^{-\frac{1}{2}} \left| \alpha_1 + \alpha_2 x_i + \alpha_3 y_i - z_i \right|. \qquad (8\text{-}1)$$

This is a discrete *nonlinear* minimax approximation problem. One
of the methods mentioned in Section 6 yields a set of x_i, y_i, z_i
points corrected to the least squares mean plane, i.e., *their*
least squares planar approximation is $z = 0$. We expect therefore
that relative to this plane the solution values of α_2 and α_3
in (8-1) will be small in magnitude in comparison with unity, and
hence that $\alpha_2^2 \ll 1$ and $\alpha_3^2 \ll 1$. Thus it is reasonable to replace
(8-1) by the problem

$$\min_{\alpha} \max_{i} \left| \alpha_1 + \alpha_2 x_i + \alpha_3 y_i - z_i \right|, \qquad (8\text{-}2)$$

which is a discrete *linear* minimax approximation problem. As
such, the problem can be regarded as a special case of the
general linear minimax approximation problem

$$\min_{\alpha} \left\| e(\alpha) \right\|, \quad e(\alpha) = z - A\alpha, \quad \left\| e \right\| = \max_{1 \leqslant i \leqslant m} \left| e_i \right|, \qquad (8\text{-}3)$$

where A is a given $m \times w$ matrix, for some w, z a given
m-element vector, and α a w-element vector to be determined.

Good algorithms exist for solving such problems. One that
we have used successfully is based upon the following linear
programming formulation (see Watson [1980, p.39], for example)
of (8-3):

minimize β

subject to $-\beta \leqslant e_i \leqslant \beta$, $i = 1, 2, \ldots, m$,

or

minimize β

subject to $\begin{pmatrix} A & g \\ -A & -g \end{pmatrix} \begin{pmatrix} \alpha \\ \beta \end{pmatrix} \geqslant \begin{pmatrix} z \\ -z \end{pmatrix}$,

where $g = (1, 1, \ldots, 1)^T$. This is a linear programming problem
in the $w + 1$ variables α, β. It is not, however, in a form suitable for the direct application of standard techniques, since

the elements of α are not constrained to be non-negative.   However, the dual formulation of the problem is

$$\text{maximize } (u - v)^T z$$
$$\text{subject to } A^T(u - v) = 0 ,$$
$$g^T(u + v) \leqslant 1 , \qquad\qquad (8\text{-}4)$$
$$u , v \geqslant 0 ,$$

which has the advantages that (a) all variables are constrained
to be non-negative, (b) there is only one inequality constraint,
and thus just one slack variable is required, and (c) the number
of equations is $w + 1$ (four for the plane separation problem)
instead of $2m$.

Barrodale & Phillips [1975] give a very efficient algorithm for solving this dual formulation. Their algorithm, which
is a version of the standard simplex method modified to take
advantage of the special structure of (8-4), is used in FLATPAK,
for which application it converges in very few iterations.

In cases where it appears unreasonable to replace formulation (8-1) by (8-2), an iterative method for solving the full
discrete nonlinear minimax problem may be employed. Watson [1980,
Chapter 10] refers to a number of such methods, each iteration of
which constitutes the solution of a linear subproblem, followed
by a linear search in the space of the variables. Each subproblem can be cast as a linear program of the form (8-4), and thus
the efficient method already mentioned can be used. (This
refinement is not incorporated in the current release of FLATPAK.)

9.   SOFTWARE IMPLEMENTATION

The software comprising FLATPAK is written in ANSI
Standard Fortran (ANSI [1966]) and is accompanied by full user
documentation (Compeda [1980]). The code is modularized, consisting of a main program and 48 subroutines. Of the subroutines,
18 were specially designed for FLATPAK, 20 are 'generic' mathematical modules, and the remaining 10 provide graphical output.

The decision to make explicit a clear distinction between application, mathematical and graphics routines was made for a number of reasons. First, the mathematical modules had been well-tested in another environment, being part of a pre-release version of DASL — the NPL Data Approximation Subroutine Library (Anthony *et al.* [1982]). Second, the graphical routines were derived from established software from the NPL Algorithms Library (Cooper [1979]), and incorporate techniques based upon those given by Heap [1972] with which much experience had earlier been gained. Third, the 'applications' part of the package is itself modularized so as to permit considerable flexibility in future releases of FLATPAK. The current release provides facilities for a triangular measurement grid and a gradient-measuring instrument. Of those 18 'applications' routines, six relate to the grid, and seven also to the measuring instrument (four of which are simple data checking and array dimensioning routines). The parameter interface between routines is such that if a version of the package were required for a different grid or a new measuring instrument, only a relatively small proportion of the total package would require modification, by replacing appropriate routines.

ACKNOWLEDGEMENT

I am extremely grateful to G.T. Anthony and K.G. Birch for their valuable comments.

REFERENCES

ANSI 1966, USA Standard FORTRAN. *American National Standards Institute* Publication X 3.9 — 1966.

Anthony, G.T., Cox, M.G. & Hayes, J.G. 1982, An outline of the NPL Data Approximation Subroutine Library. *National Physical Laboratory* Document DASLDOC1.

Barrodale, I. & Phillips, C. 1975, Algorithm 495: solution of an overdetermined system of linear equations in the Chebyshev norm. *ACM Trans. Math. Software* 1, 264-270.

Birch, K.G. & Cox, M.G. 1973, Calculation of the flatness of
    surfaces; a least squares approach. *National Physical
    Laboratory* MOM Report No. 5.

Birch, K.G. & Harrison, M.R. 1974, Calculation of the flatness
    of surfaces; an addendum to NPL Report MOM 5. *National
    Physical Laboratory* MOM Report No. 9.

BSI 1972 Specification for surface plates and tables. *British
    Standards Institution*, BS 817.

COMPEDA 1980, FLATPAK User Guide. *Compeda publication* D A59 060.

Cooper, J.R.A. 1979, NPL Algorithms Brief Guide — 1979.
    National Physical Laboratory CSU Report No. 3/79.

Cox, M.G. 1981, The least squares solution of overdetermined
    linear equations having band or augmented band structure.
    IMA J. Numer. Anal. 1, 3-22.

Cox, M.G. & Jackson, K. 1983, Algorithms and software for
    engineering metrology: a statement of need. *National
    Physical Labortatory* DMOM Report No. 65.

Duff, I.S. & Reid, J.K. 1976, A comparison of some methods fo
    the solution of sparse over-determined systems of linear
    equations. *J. Inst. Math. Appl.* 17, 267-280.

Gentleman, W.M. 1974a, Basic procedures for large, sparse or
    weighted linear least squares problems. *Appl. Statist.* 23,
    448-454.

Gentleman, W.M. 1974b, Regression and the QR decomposition.
    Bull. IMA 10, 195-197.

Gill, P.E., Golub, G.H., Murray, W. & Saunders, M.A. 1974,
    Methods for modifying matrix factorizations. *Math. Comput.*
    23, 505-535.

Golub, G. 1965, Numerical methods for solving linear least
    squares problems. *Numer. Math.* 7, 206-216.

Heap, B.R. 1972, Algorithms for the production of contour
    maps over an irregular triangular mesh. *National Physical
    Laboratory* NAC Report No. 10.

Manneback, P.E. 1983, A direct approach for column ordering and
    symbolic factorization in large sparse least squares problems.
    Dept. of Maths., Faculties Univ. de Namur, Belgium, Report No.
    83/11.

Peters, G. & Wilkinson, J.H. 1970, The least squares problem
    and pseudo-inverses. *Comput. J.* 13, 309-316.

Watson, G.A. 1980, *Approximation Theory and Methods*, Chichester,
    Wiley.

2

A simple new design formula for chain links

G. T. ANTHONY, T. A. E. GORLEY, AND J. G. HAYES

1.   INTRODUCTION

British Standards for lifting gear were first discussed
some forty years ago. Various standards have since been drawn
up which specify the dimensions of components, the materials
from which they must be made, the tests they must undergo before
they can be given a certificate and all other information necessary to make them as safe as possible. A continual watch is kept
by the relevant British Standards Institution committees on
improvements in design, materials and manufacturing techniques
and on changes in requirements. When necessary, modifications
to the standards are made.

Sometimes, 'special' components are manufactured which
are 'one-off' designs and do not therefore warrant such modifications. These components comply with the standards as far as
general shape and materials are concerned, but not with the
standard set of dimensions. In order that such designs may be
described as "Special ... to BS ...", it has been British
Standard practice to include a design formula in the standards
wherever possible.

For those of the early standards which specify chain
links, a formula was developed at NPL which, given a link's
dimensions and design stress, determined the design load. The
inclusion of this formula has been generally regarded with much

25

favour by those involved in the manufacture, usage and certifi-
cation of the links. The formula has stood the test of time and
has, by an increase in the design stress, been the basis for the
design of links made of the new, stronger, alloy steels. The formula is applicable to *all* chain links; it embraces the links
that make up a length of chain, as well as the master links,
joining links and intermediate links which connect lengths of
chain in chain slings (Figure 1 shows a typical four-leg sling).

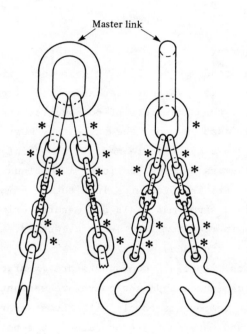

Master link

✱ Intermediate and joining links

Figure 1   A typical four-leg sling

During recent years, International Standards for lifting
gear have been discussed. Progress has been varied; some standards have been published already, while others have been delayed
for want of agreement. One of those recently published is the
standard for welded chain slings (ISO 4778 and BS6304). A

major area of difficulty in drafting this standard was in deciding the requirements for master links and intermediate links.
All delegations to ISO (International Standards Organization)
agreed on the necessity to specify the basic mechanical properties of the link (e.g. minimum failure loads and proof loads);
but some also wished to specify the links completely, with all
the dimensions given; some wished to see also the performance
of these links completely specified, leaving the manufacturer to
choose appropriate dimensions; others were not in favour of the
latter course, nor did they insist on the former — they considered it satisfactory to specify the *means* of obtaining the link
dimensions. The UK delegation was in the last category and,
accordingly, they proposed the formula contained in British
Standards as a means of determining the dimensions.

After it had been proposed, the formula was criticised on
two counts — first, that it was not convenient to use; second,
that the design stresses were unrealistic. To cover the latter
objection, tests were undertaken at the Safety in Mines Research
Establishment, Sheffield, to demonstrate that the design stress
does have a physical meaning. The first objection required that
the formula should express the diameter of the material from
which the link is to be made directly in terms of the inside
dimensions of the link, the design stress and the design load.
Unfortunately, the methods on which the stress analysis of a
chain link is based do not enable a formula of the desired form
to be obtained directly. Therefore it was decided to try to
construct an acceptable formula purely by numerical methods.
First, the stress analysis was used to obtain 'data' for a range
of shapes and sizes of link; then, formulae were devised which
'fit' this data to acceptable accuracy,

The purpose of this chapter is to describe the various
steps of the numerical investigation which ultimately yielded
these new formulae; not only the successful steps but also
those which failed and which are perhaps the most illuminating.

2.   THE OLD DESIGN FORMULA

The formula for the design of special links that has been
included in British Standards for about forty years is defined
in terms of the notation shown in Figure 2.

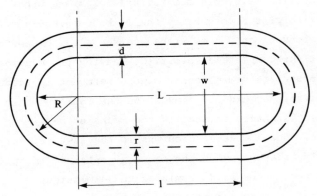

Figure 2   A chain link

The following simple identities exist:
$$d = 2r , \quad w = 2R - 2r , \quad L = l + 2R - 2r. \tag{2-1}$$
For values of $L \geqslant 2.55d$, the formula is:

$$F = \frac{0.224\,fd^3}{w + 0.4d} \left(1.75 + \frac{w+d}{L+d}\right) \tag{2-2}$$

in which the units of F, f, w, L and d are self-consistent and

F = design load,

d = nominal diameter of the material from which the
link is made,

L = internal length of the link,

w = internal breadth of the link,

f = design stress: the maximum nominal tensile stress
in the component when it is subject to the design
load F.

The tensile stress f to which (2-2) relates is that set
up in the extreme fibres at the extrados in the line of the load.
This is generally the maximum tensile stress in the link. When
it is not, the maximum occurs in the extreme fibres at the intra-
dos where the straight and circular portions of the link join.

In order to cater for this alternative, the formula for F must be multiplied by the reduction factor $0.22(2+L/d)$ whenever $L < 2.55\,d$. In this way, the difference between the results, as produced from the theoretical analysis and from the formula, are less than 4% in practice and are generally less than 2%.

Unfortunately, the complete derivation of (2-2) has been lost. However, some clues to its derivation are confirmed in Gough, Cox & Sopwith [1934] and it will be shown in Section 6, when results are compared, that its accuracy is adequate for design purposes.

3.   THE STRESS IN CHAIN LINKS

Our need, then, is for a formula expressing d directly in terms of the other parameters in (2-2). To provide a sound basis for this purpose, T.A.E. Gorley (see Anthony, Gorley & Hayes [1977]) derived the following formula by linear elastic stress analysis:

$$f \;=\; \frac{F}{2\pi r^{2}}\left(\frac{l + r^{2}/2ZR}{l + \pi r^{2}(Z+1)/4ZR}\right)\!\left(1 + \frac{r}{Z(R+r)}\right) \quad (3\text{-}1)$$

or

$$f \;=\; \frac{F}{2\pi r^{2}}\left(\frac{l + r^{2}/2ZR}{l + \pi r^{2}(Z+1)/4ZR}\right)\!\left(1 - \frac{r}{Z(R-r)}\right) + \frac{F}{2\pi r Z(R-r)} \quad (3\text{-}2)$$

whichever is the larger. Here Z is given by

$$Z \;=\; (r/R)^{2}\left\{\left(1-(r/R)^{2}\right)^{\frac{1}{2}} + 1\right\}^{2}. \quad (3\text{-}3)$$

That there are two forms of f ((3-1) and (3-2)) reflects the fact that the maximum stress can occur at either of two different positions on the link depending on its geometry.

{Note that equations (2-1) relate the above parameters to those in (2-2), the parameters used in practice.}

4.   THE CRITERIA FOR A NEW FORMULA

It is reasonably clear that the Old Design Formula (2-2) could not be rewritten to give d directly in terms of its other parameters.

It was therefore necessary to devise some other means of

obtaining an acceptable formula. In fact data from equations
(3-1) and (3-2) of the stress analysis were used to generate
self-consistent combinations of F, f, d, L and w for all likely
values of link dimensions, and data-fitting techniques then used
to construct a formula for d in terms of F, f, w and L. Practical limits of link dimensions were taken to be $0.1 \leqslant r/R \leqslant 0.6$
and $0 \leqslant l/R \leqslant 4.0$.

It was decided to maintain, and if possible improve upon,
the accuracy of the Old Design Formula while aiming to achieve
as simple a formula as possible. Designers will often be working to tight schedules and a complicated formula may increase
the likelihood of errors being made for very little improvement
in accuracy. Also, it has to be remembered that the diameter of
the steel rod, from which the links are made, is allowed to
differ from the specified diameter by a given tolerance. For
example, BS 4942: Part 2: 1973, Grade M non-calibrated short link
chain for lifting purposes, specifies a tolerance of +2% to -6%
on chain of nominal diameter up to and including 16 mm, and a
tolerance of ±5% on chain of nominal diameter equal to or greater than 18 mm.

5.   DERIVATION OF A NEW FORMULA

5.1   *Choice of Working Parameters*

Since we wish to devise the simplest possible formula, we
must devote particular attention to the choice of the form in
which to consider the design parameters F, f, w, L and d. First
of all we shall seek some guidance from the Old Design Formula
(2-2), since this formula is easier to comprehend than the more
precise formulae (3-1) and (3-2). We observe that the number of
parameters in (2-2) can be reduced from 5 to 3 simply by con-
verting to non-dimensional parameters. For example, if we write

$$x_1 = (F/fL^2)^{\frac{1}{3}}, \quad y_1 = w/L, \quad z_1 = d/L \qquad (5-1)$$

formula (2-2) becomes

$$x_1^3 = \frac{0.224 z_1^3}{y_1 + 0.4 z_1} \left(1.75 + \frac{y_1 + z_1}{1 + z_1} \right) \qquad (5\text{-}2)$$

and the right-hand side has to be multiplied by $0.22(2 + 1/z_1)$ when $L < 2.55\,d$. This suggests, since we wish to express d in terms of the other design parameters, that it might be fruitful to seek a formula relating z_1 to x_1 and y_1, and certainly the search is likely to be simpler with only three parameters to consider rather than five. The cube root has been included in the definition of x_1 because then the dominant terms x_1^3 and z_1^3 in (5-2) occur with the same power, and so we might expect our formula to consist mainly of a linear relationship between z_1 and x_1 (or indeed z_1^3 and x_1^3 or any other power).

We used the approximate formula (2-2) to suggest the use of the parameters x_1, y_1 and z_1 given in (5-1). However, we can also see that formulae (3-1) and (3-2) can be expressed entirely in terms of the three non-dimensional parameters F/fr^2, l/R and r/R. These in turn can, by means of equations (2-1), be expressed entirely in terms of the parameters of (5-1), and so it follows that equations (3-1) and (3-2) can, like equation (2-2), be rewritten in terms of the parameters x_1, y_1 and z_1 of (5-1).

Figure 3 shows the relationship between z_1 and x_1 for different values of y_1, as computed from formulae (3-1) and (3-2), with f equated to the larger of these. The slope of each of these curves changes only slowly, as we expected, except of course for the discontinuous change at the point where equation (3-2) takes over from (3-1). For each value of y_1, the unbroken curve extends over the range of z_1 which corresponds to the practical ranges we selected for r/R and l/R, namely 0.1 to 0.6 for r/R and 0.0 to 4.0 for l/R. The combinations of z_1 and z_2 which are compatible with these latter ranges are indicated in Figure 4: any combination is compatible which, when plotted on the diagram, lies within the quadrilateral shown. It

G.T. ANTHONY *et al.*

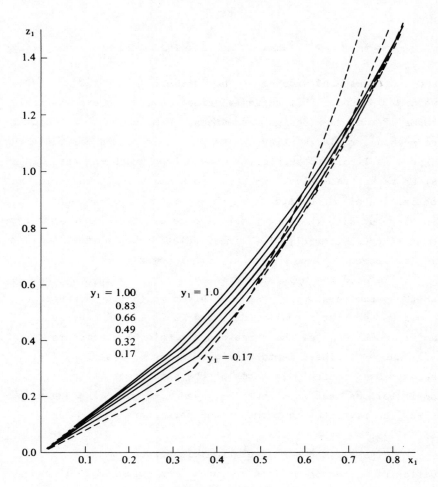

Figure 3

$$z_1 = \frac{d}{L} \qquad y_1 = \frac{w}{L} \qquad x_1 = \left(\frac{F}{fL^2}\right)^{\frac{1}{3}}$$

Unbroken curves are for ranges of z_1 in the quadrilateral of Figure 4 (for $y_1 = 0.17$, there is only a short section, near $x_1 = 0.3$). Broken curves extend the range to

$$0.03 \leqslant z_1 \leqslant 1.5$$

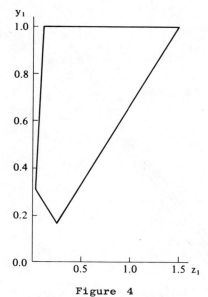

Figure 4

$$y_1 = w/L \qquad z_1 = d/L$$

Quadrilateral contains combinations of y_1 and z_1 corresponding to $0.1 \leqslant \dfrac{r}{R} \leqslant 0.6$ and $0 \leqslant \dfrac{l}{R} \leqslant 4$

is, then, the full curves of Figure 3, taken together, for which we wish to obtain an approximating formula. When obtained, the formula should not be used outside the ranges on which it has been based and validated, so strictly the user should check that his particular values of z_1 and y_1 lie within the quadrilateral in Figure 4. Clearly it would be more convenient if we could devise a formula which was valid over the full range of z_1 (0.03 to 1.50) for all values of y_1: the user would then simply need to check that z_1 and y_1 each lay between specified numerical limits. Of course, this extension would only be worthwhile if it did not require a more complicated formula. The broken portions in Figure 3 show the extensions of the curve to the full range of z_1. They suggest that a formula to cover them would indeed need to be more complicated, but we shall see later a situation in which this is not the case.

Another choice of working parameters which equation (5-2) suggests should be considered, is $\log z_1$ and $\log x_1$. The use of the logarithm involves in itself some computational complication in a formula, but on the other hand it might result in compensating simplifications in the form of the formula required. The relationship between $\log z_1$ and $\log x_1$, for various values of y_1, is shown in Figure 5.

Furthermore, the set of parameters in (5-1) is not the only set of three in which equations (3-1), (3-2) and (2-2) can

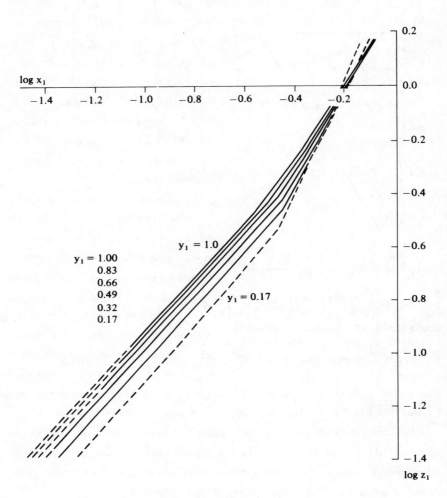

Figure 5

$$z_1 = \frac{d}{L} \qquad y_1 = \frac{w}{L} \qquad x_1 = \left(\frac{F}{fL^2}\right)^{\frac{1}{3}}$$

Unbroken curves are for ranges of z_1 in the quadrilateral of Figure 4 (for $y_1 = 0.17$, there is only a short section near $\log x_1 = -0.5$). Broken curves extend the range to
$$0.03 \leqslant z_1 \leqslant 1.5$$

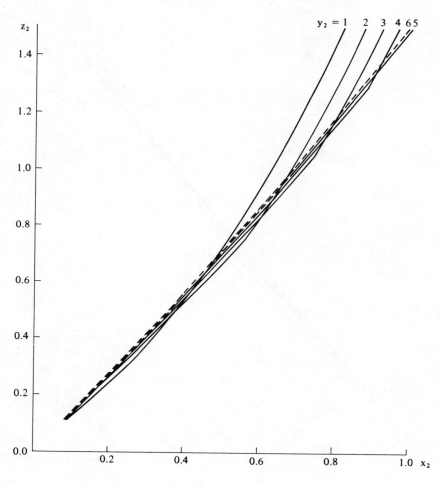

Figure 6

$$z_2 = \frac{d}{w} \qquad y_2 = \frac{L}{w} \qquad x_2 = \left(\frac{F}{fw^2}\right)^{\frac{1}{3}}$$

Unbroken curves are for ranges of z_2 in the trapezium of Figure 8 (for $y_2 = 6$, there is only one point, at the extreme right). Broken curves extend the range to
$$0.1 \leqslant z_2 \leqslant 1.5$$

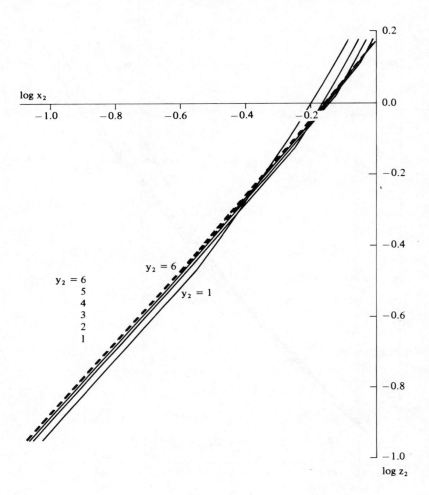

Figure 7

$$z_2 = \frac{d}{w} \qquad y_2 = \frac{L}{w} \qquad x_2 = \left(\frac{F}{fw^2}\right)^{\frac{1}{3}}$$

Unbroken curves are for ranges of z_2 in the trapezium of Figure 8 (for $y_2 = 6$, there is only one point, at the extreme right). Broken curves extend the range to
$$0.1 \leqslant z_2 \leqslant 1.5$$

be expressed. We must avoid any set for which d occurs in more than one parameter, since it is d we wish to isolate, but there remains the set corresponding to (5-1) with L and w interchanged, namely

$$x_2 = (F/fw^2)^{\frac{1}{3}} \quad y_2 = L/w \quad z_2 = d/w . \quad (5-3)$$

The graphs of z_2 against x_2 and $\log z_2$ against $\log x_2$ are shown for various values of y_2 in Figures 6 and 7. The combinations of values of y_2 and z_2 corresponding to the restrictions on r/R and l/R are shown in Figure 8. The broken portions in Figures 6 and 7 show the extensions of the curves to the full range (0.1 to 1.5) of z_2, and here we see a distinct possibility that the extensions will not require a more complicated formula than the unbroken curves alone.

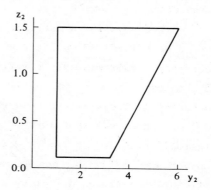

Figure 8

$$y_2 = L/w \quad z_2 = d/w$$

Trapezium contains combinations of y_2 and z_2 corresponding to $0.1 \leqslant r/R \leqslant 0.6$ and $0 \leqslant l/R \leqslant 4$

5.2 *Search for a Single Formula*

Essentially, however, we wish to find a simple formula representing one or other of the four relationships shown by the sets of unbroken curves in Figures 3, 5, 6 and 7, to an accuracy of perhaps $\pm 2\%$. Taking Figure 3 as an example, we may consider our problem as being one of expressing z_1 as a function of x_1 (representing a typical curve in the diagram) in which the coefficients are functions of y_1. Our first aim was to investigate whether a single formula would be adequate for one or other of the four relationships, despite the presence of slope discontinuities. However, in Figures 5, 6 and 7, we see that a typical curve consists of a long, almost straight, section followed by a curved section departing appreciably from it. We cannot expect a simple formula, such as a quadratic, for example, to follow such a shape at all precisely. In Figure 3, on the other hand, the straight sections are much shorter, and this relationship appears to offer hope for a single formula. The fact that the slope discontinuities occur at closely similar values of x_1 is also encouraging.

For devising a mathematical form for the relationship, it would be helpful if we had a diagram in which the size of the accuracy requirement was clearly visible everywhere. In Figure 3, basically because z_1 varies by a factor of more than 40, 2% accuracy is too small to be distinguished at the smaller values, and we cannot see what sort of formula would be adequate in this region. However, it may be seen from the figure that all the curves can be contained between a pair of lines through the origin whose slopes differ by a factor of only about 3. Thus values of z_1/x_1 will vary by a factor of no more than 3, and if we plot z_1/x_1 against x_1 so that the range of z_1/x_1 values span 20 cms on the diagram, a 2% error will correspond to 2 mm at the least. We can go further, however. Since in Figure 3 the curve segments to the left of the discontinuities approximate to straight lines through the origin, the major part of the effect of y_1 on the whole 'left-hand' relationship can be uncovered by examining algebraically the behaviour of formula (3-1) close to the origin. In fact, from (3-1), using (2-1), we find that for small values of z_1 we have the approximate relationship

$$z_1 = x_1 g_1(y_1) \qquad\qquad (5-4)$$

where

$$g_1(y_1) = 2/(\pi(\tfrac{1}{2}\pi - 1 + 1/y_1))^{\frac{1}{3}} .$$

Using this result, we can then plot $z_1/x_1 g_1(y_1)$ against x_1, but in fact it turns out to be better to plot the quantity

$$P = z_1/x_1 - g_1(y_1) . \qquad\qquad (5-5)$$

This we have done in Figure 9, but truncating the values of P at the highest value which occurs on the unbroken curves. At the

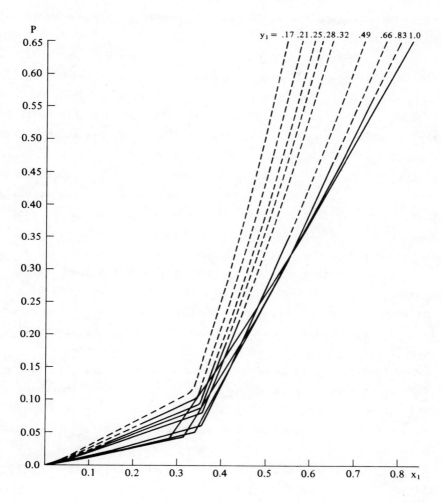

$$y_1 = .17\ .21\ .25\ .28\ .32\quad .49\quad .66\ .83\ 1.0$$

Figure 9

$$P = \frac{z_1}{x_1} - g_1(y_1) \qquad z_1 = \frac{d}{L} \qquad y_1 = \frac{w}{L} \qquad x_1 = \left(\frac{F}{fL^2}\right)^{\frac{1}{3}}$$

$$g_1(y_1) = 2/\left[\pi\left(\tfrac{1}{2}\pi - 1 + 1/y_1\right)\right]^{\frac{1}{3}}$$

Unbroken curves are for ranges of z_1 in the quadrilateral of Figure 4 (for $y_1 = 0.17$, there is only a short section near $x_1 = 0.3$). Broken curves extend the range to

$$0 \leqslant P \leqslant 0.65$$

same time, some additional values of y_1 have been included so as
to exhibit the relationship more fully in the critical region of
the discontinuities. Of course, plotting P rather than z_1/x_1
simply subtracts a different constant from each of the curves of
z_1/x_1 against x_1, so that all the curves, when extended, pass
through the origin, as is shown in the diagram. We observe that
the variation due to y_1 which remains in P is, for the unbroken
curves, less than 0.1 for all values of x_1. Thus, since $g_1(y_1)$
varies in value from 0.7 to 1.2, we have succeeded in removing
most of the effects of y_1 on z_1/x_1, for the right-hand as well
as the left-hand relationship. Corresponding to an accuracy of
$\pm 2\%$ in z_1, the (absolute) accuracy required in approximating P
is $0.02\, z_1/x_1 = 0.02\, (P + g_1(y_1))$, from (5-5). Values of P read
from Figure 9 then show that the accuracy required in P varies
from about ± 0.015 (for $y_1 = 0.17$) to ± 0.025 (for $y_1 = 1.0$) at
the lowest values of x_1, and from about ± 0.025 to ± 0.035 similarly at the highest values. The lowest of these requirements
(0.015) corresponds to $4\frac{1}{2}$ mm on the diagram.

However, it is now clear that, because of the size of the
slope discontinuities, a simple formula cannot approximate the
full extent of these curves at all accurately. We have already
reached the same conclusion in respect of Figures 5, 6 and 7, and
so we abandon the attempt to find a single formula and seek instead a pair of simple formulae, one for the curves on the left
and one for those on the right.

5.3   *Search for a Pair of Formulae*

There are two different strategies we might adopt when
seeking a pair of formulae. We might split the range of x_1 at
some value about 0.35 and devise one formula for the curves to
the left of this value and another for those to the right: or
we might devise one formula for the curves to the left of their
respective discontinuities, and another to the right. In the
latter case, when it comes to practical use of the formulae, the

intention would be to evaluate d from both formulae and take
the larger of these values. Thus the first strategy has the
advantage that only one formula need be evaluated for each particular practical application. However, if we try to devise a
pair of formulae appropriate to this strategy, a careful examination of Figure 9 indicates that we are unlikely to do better than
to have a single straight-line approximation for the curves on
the left, and another one on the right, independent of y_1. This
would give a maximum error of perhaps three or four percent.

   We shall return to this case in Section 6, but in the
meantime we shall consider the derivation of a pair of formulae
appropriate to the second strategy. In this case, our initial
reasons for preferring the relationship of Figure 3 (and thus
Figure 9) are no longer valid, and if we re-examine Figures 3
and 6 from the new point of view, we see that the curves in the
latter are the more promising. The long straight left-hand segments are now an advantage and their variation with y_2 is
already within about $\pm 5\%$. Moreover, the slowly-curving right-
hand segments are, to reasonable percentage accuracy, not far
from being parallel and equi-spaced, so that we can hope that a
formula of the form

$$z = a_1 + b_1 x + c_1 x^2 + d_1 y \qquad (5\text{-}6)$$

might be adequate. Here, a_1, b_1, c_1 and d_1 are constants, and
x, y and z are the working parameters defined in (5-3), the
suffix 2 having been dropped as unnecessary since for most of
the rest of the chapter we shall be confining our attention
solely to these parameters. If the form (5-6) proves inadequate,
we may consider adding to it a term in xy (to allow a degree of
non-parallelism) and/or one in y^2 (to allow a degree of unequal
spacing).

   To devise a corresponding trial form for the left-hand
segments, we need, as with Figure 3, to be able to view these
segments with greater accuracy than is possible with Figure 6.
We can show, corresponding to equation (5-4), that for small

values of z we have the approximate relationship

$$z = xg(y) \qquad (5-7)$$

where

$$g(y) = 2y^{\frac{1}{3}}/(\pi(\tfrac{1}{2}\pi - 1 + y))^{\frac{1}{3}} \qquad (5-8)$$

and values of $g(y)$ lie between 1.18 and 1.33. In Figure 10, therefore, we have plotted against x the quantity

$$Q = z/x - g(y) , \qquad (5-9)$$

restricting its range to the lower values, since we are concerned here with only the left-hand segments. Corresponding to an accuracy of $\pm 2\%$ in z, the absolute accuracy required in approximating Q can be shown, as in the discussion of equation (5-5), to vary from ± 0.025 at low values of x to ± 0.03 at the higher values in the diagram.

We can see that a single straight line of the form

$$Q = a_2 + b_2 x , \qquad (5-10)$$

coinciding with the straight section of the curve for $y = 4$, will approximate all the segments, including their extensions, with a maximum error of little more than 1%. Indeed, omitting the constant term from (5-10) would not greatly increase the maximum error, though the errors at the left-hand end of the curves would increase. Moreover, since the spacing of adjacent segments, except the segment for $y = 1$, is almost independent of y and proportional to x, a term in xy added to (5-10) would account for most of the variation with y in the diagram. Nevertheless, equation (5-10), without its constant term, is accurate enough for our purpose, and so, substituting for Q from (5-9), we have the trial form

$$z/x = b_2 x + g(y). \qquad (5-11)$$

However, we can show that $g(y)$, defined in (5-8), can be closely approximated by a simple expression of the form $\alpha + \beta/y$, and substituting this approximation into (5-11) affects the value of z by no more than 0.3%. We then have the form

$$z/x = \alpha + \beta/y + b_2 x , \qquad (5-12)$$

where α and β may be determined first by approximating $g(y)$

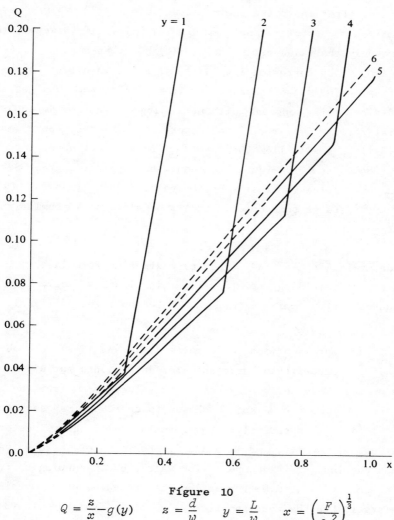

Figure 10

$$Q = \frac{z}{x} - g(y) \qquad z = \frac{d}{w} \qquad y = \frac{L}{w} \qquad x = \left(\frac{F}{fw^2}\right)^{\frac{1}{3}}$$

$$g(y) = 2y^{\frac{1}{3}} / \left[\pi\left(\tfrac{1}{2}\pi - 1 + y\right)\right]^{\frac{1}{3}}$$

Unbroken curves are for ranges of z in trapezium of Figure 8 (for $y = 6$, there is only one point, at the extreme right). Broken curves extend the range to $0 \leqslant x \leqslant 1.5$. The figure is, however, truncated at $Q = 0.2$.

and then, after inserting their values into (5-12), b_2 may be determined by approximating the formula (3-1). If, however, we insert only the value of β into (5-12) and then obtain α, as well as b_2, by approximating (3-1), we shall in effect be incorporating in α the constant a_2 from (5-10), and thus get the benefit of this constant without adding another term to our initial form. Similarly, if we determine β also by approximating (3-1) we shall allow the term β/y to take out some of the variation with y in Figure 10, though perhaps not a great deal as this variation is clearly better related to y than to $1/y$. However, there is some benefit and no penalty, so we shall write our trial form as

$$z = x(a_2 + b_2 x + c_2/y) \qquad (5-13)$$

and determine a_2, b_2 and c_2 by approximating formula (3-1). Most of the remaining variation in y in Figure 10 could be accounted for by adding a term $d_2 x^2 y$ to the right of (5-13).

5.4   *Selection of Data Points*

The approximations to formulae (3-1) and (3-2) were obtained by a data-fitting process. The data points for use in this process were derived by computing values of x from the formulae at values of y and z chosen to cover the ranges 1 to 6 and 0.1 to 1.5 respectively. For formula (3-1), the values of y used were 1, 1.6, 1.8, 2, 3, 4, 5 and 6. For each of these values except the last, x was computed for values of z running up from 0.1 at intervals of 0.05 for as long as the value of x obtained was larger than the corresponding value of x from formula (3-2). In the case of $y = 6$, the crossover point of the two formulae occurs beyond $z = 1.5$, so 1.5 was the largest value used. This selection of data points is somewhat arbitrary, but the aim is simply to cover the behaviour of the formula adequately. The crossover points themselves were also included in the data, since it is advisable to cover the full relevant range of the formula. It was for the same reason that the extra values of y between

1 and 2 were included, since, as Figure 10 indicates, there are
values of y in this region for which the associated curves lie
partially below any of those in the figure.

For formula (3-2), the values of y used ran from 1.0 at
intervals of 0.5 up to 5.0. For each of these values, x was computed for values of z running down from 1.5 at intervals of 0.2
for as long as formula (3-2) provided a value of x greater than
formula (3-1). Again the crossover points themselves were also
included. The crossover points for $y = 5.5$ and 6.0 occur beyond
$z = 1.5$, and so these values of y are not included in the data
for formula (3-2).

5.5   *Method of Fitting*

We now wish to fit to the two sets of data just described
the appropriate trial form (5-13) or (5-6), that is, to derive
for the coefficients in the forms those values which give the
best fit according to some criterion. There is a choice of
criteria, but with any choice, since it is percentage accuracy
in z which is of interest, we shall be concerned with the set
of percentage residuals ε_i, with $i = 1$ to N, where

$$\varepsilon_i = 100(z_i - z_i^*)/z_i . \qquad (5-14)$$

Here we have assumed the data points to be numbered in
some order from 1 to N, where N is the relevant number of
data points. The quantity z is the data value of z for the ith
data point, and z_i^* is the corresponding value of z obtained by
substituting x_i and y_i into (5-13) or (5-6) as appropriate.
Probably the most suitable fitting criterion for our purpose is
the minimax criterion, which chooses those values of the coefficients which make the largest of the ε_i (regardless of sign) as
small as possible. In other words, the maximum percentage error
is as small as it can be for the chosen form of formula. However,
since the choice of criterion is not likely to be critical and
since an appropriate computer algorithm was more readily available, the least-squares criterion (minimizing the sum of squares

of the ε_i) was used to obtain our first fits. The minimax criterion was used for the final fits.

The algorithm using the least-squares criterion was written by M.G. Cox at NPL; that using the minimax criterion was taken from Barrodale & Young [1966].

5.6   *Results*

The least-squares fit of the form (5-13) to the data from formula (3-1) is

$$z = x(1.349 + 0.181\,x - 0.205/y)\,. \qquad (5\text{-}15)$$

The root mean square residual for this fit is 0.7%, and the maximum residual is 2.2%. The root mean square residual for the form (5-13) with an additional term in x^2y is 0.2%, and the maximum residual is 0.8%, indicating as expected that this term explains most of the remaining variation. However, (5-15) satisfies our accuracy requirement and so there is little point in complicating our formula for the sake of the extra accuracy.

The least-squares fit of the form (5-6) to the data from formula (3-2) is

$$z = 0.091 + 1.029\,x + 0.947\,x^2 - 0.117\,y\,. \qquad (5\text{-}16)$$

The root mean square residual is 0.8%, and the maximum residual 2.5%. Adding to (5-6) a term in either xy or y^2 does not improve the fit appreciably, but adding both terms reduces the root mean square residual to 0.2% and the maximum residual to 0.5%. Again, however, the formula (5-16) is of acceptable accuracy, and there is little point in accepting the complication of two extra terms.

The maximum residuals are, of course, reduced if the minimax criterion instead of the least-squares one is used for fitting to the data. The minimax solutions corresponding to the solutions (5-15) and (5-16) are respectively

$$z = x(1.332 + 0.188\,x - 0.179/y) \qquad (5\text{-}17)$$

and

$$z = 0.097 + 1.019\,x + 0.977\,x^2 - 0.121\,y\,. \qquad (5\text{-}18)$$

The maximum residuals for these two formulae are respectively 1.3% (cf. 2.2% for least-squares) and 1.6% (cf. 2.5%).

We may note that these values are well within our accuracy
requirement.

However, because of the particular values of some of the
coefficients in (5-17) and (5-18), it was apparent that an appropriate rounding of them would result in a significant simplification in the arithmetic of evaluating the formulae, hopefully
without increasing the maximum residuals beyond acceptable bounds.
In fact, if we replace (5-17) and (5-18) respectively by

$$z = 0.2\,x(6.7 + x - 1/y) \qquad (5\text{-}19)$$

and

$$z = 0.1 + x(1 + x) - 0.12\,y \qquad (5\text{-}20)$$

the corresponding maximum residuals are 2.4% and 2.1%. These
were considered acceptable, and so finally, with self-consistent
units for the basic parameters F, f, d, L and w, our new
design formula is

> $d/w = 0.2\,x(6.7 + x - w/L)$
> or $\quad 0.1 + x(1 + x) - 0.12\,L/w$,
> whichever is the larger.
>
> Here, $x = (F/fw^2)^{\frac{1}{3}}$, and the
> formula is valid in the ranges
> $1 \leqslant L/w \leqslant 6$ and $0.1 \leqslant d/w \leqslant 1.5$.

$\qquad (5\text{-}21)$

The error in this formula, relative to formulae (3-1) and
(3-2), is shown graphically in Figure 11. Specifically, to obtain the error of a particular value of d/w and of L/w, these
values were first substituted, via (2-1), into both formulae (3-1)
and (3-2) to obtain a pair of x values. The large of these, and
the given value of L/w, were substituted into formula (5-21) to
obtain the corresponding approximate value of d/w. Comparison
of this value with the original value of d/w then gives its
percentage error, and therefore the percentage error of d. The
presence of two slope discontinuities for most values of L/w is
due to the fact that the crossover point of the two parts of formula (5-21) and the crossover point of the formulae (3-1) and

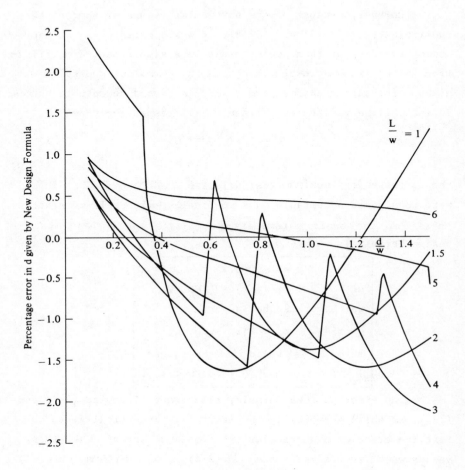

Figure 11
Percentage error in *d* given by new design formula

(3-2) occur at different values of d/w.

We observe that the error of the new formula is almost everywhere less than 1.5%, and indeed mostly less than 1%. Its maximum is 2.4%.

6.   CONCLUDING REMARKS

We have described, in Section 5, only the main steps of the investigation which led to the New Design Formula (5-21). Other possibilities, such as the relationship between z_1^3 and x_1^3, were also investigated but proved less promising than those described. Other avenues which could have been further explored, once the search for a single formula had been abandoned, were the relationships of Figures 5 and 7, but logarithms involve additional effort in the use of the derived formulae and an adequately compensatory simplification in the formulae seemed unlikely.

There remains the possibility mentioned at the end of the first paragraph of Section 5.3, to which we said we would return. This is, to split the range of x_1 in Figure 9 at some (best) point and fit a single straight line to the curves to the left of the point and another to those to the right. Doing this, using the minimax criterion, we find that the position of the dividing point is not very critical and $x_1 = 0.35$ is a satisfactory value. The maximum error of the fit is 3.1%. This is greater than that for the pair of formulae (5-15) and (5-16), namely 1.6%, and that for the rounded formulae in (5-21), namely 2.4%, but is probably still acceptable. Writing the derived pair of straight lines in terms of the working parameters we get the formulae

$$z_1 = x_1 \left(0.25\, x_1 + g_1 \left(y_1 \right) \right) , \qquad (6\text{-}1)$$

and

$$z_1 = x_1 \left(-0.30 + 1.11\, x_1 + g_1 \left(y_1 \right) \right) , \qquad (6\text{-}2)$$

where

$$z_1 = d/L , \quad x_1 = \left(F/fL^2 \right)^{\frac{1}{3}} , \quad y_1 = w/L ,$$

and

$$g_1(y_1) = \left(\frac{2.55}{0.57 + L/w}\right)^{\frac{1}{3}}.$$

Now $g_1(y_1)$ cannot be represented simply in terms of L/w as could $g(y)$ of (5-8), and so, though only one of (6-1) or (6-2) has to be evaluated in any particular application, we see that it is at least as much trouble to evaluate either one of them as it is to evaluate both parts of (5-21). Moreover, formula (6-1) and (6-2) are valid only in the quadrilateral of Figure 4, and so the validity of these formulae in a particular application is more trouble to check that is the validity of (5-21).

On these considerations, therefore, we recommend (5-21) as the New Design Formula for all chain links. The Old Design Formula was included in British Standards for chain slings and it will remain in effect until such time as the standards are revised or withdrawn. However, the New Design Formula has been included in the published International Standard for chain slings. It has also been adopted by the National Coal Board of the UK in their design document on Cage Suspension Gear and it will be proposed for inclusion in any revision of a present British Standard and in any new British Standard. In the ISO Standard for chain slings, the New Design Formula is included as a means of calculating the sections of master links and intermediate links — the chain itself must comply with certain prescribed International Standards. British Standards state that the formula included therein (presently the Old Design Formula) has to be used for the design of all links that do not comply with the standard set of dimensions.

Table 1 gives values of d/w as obtained from the three methods described in this report (the exact analysis, equations (3-1) and (3-2) — the New Design Formula, equation (5-21) — the Old Design Formula, equation (2.2)) for six values of L/w and for three values of the parameter $(F/fw^2)^{\frac{1}{3}}$. For the first and the third method, the process of obtaining d/w involved guessing

$\left[\dfrac{F}{fw^2}\right]^{\frac{1}{3}}$ L/w	0.2			0.6			0.8		
	E	NDF	ODF	E	NDF	ODF	E	NDF	ODF
1.0	0.240	0.236	0.243	0.929*	0.940	0.883	1.434*	1.420	1.269
2.0	0.256	0.256	0.258	0.822*	0.820	0.817	1.280*	1.300	1.198
3.0	0.263	0.263	0.264	0.827	0.836	0.832	1.170*	1.180	1.136
4.0	0.267	0.266	0.268	0.842	0.846	0.846	1.151	1.160	1.155
5.0	0.269	0.268	0.271	0.852	0.852	0.856	1.166	1.168	1.169
6.0	0.271	0.269	0.273	0.860	0.856	0.862	1.178	1.173	1.179

Table 1

Values of d/w obtained from the Exact Analysis (E), the New Design Formula (NDF) and the Old Design Formula (ODF). (*These are links in which the stress in the side is greater than that at the crown).

an initial value and thence improving on the guess until the
correct value was found; the second method (the New Design
Formula) gave d/w directly, of course. The values in the table
demonstrate that the New Design Formula represents the Exact
Analysis over a wider range of sizes of link than does the Old
Design Formula, and it is usually the more accurate of the two.
The New Design Formula will, therefore, very adequately meet the
needs of industry.

REFERENCES

Anthony, G.T., Gorley, T.A.E. & Hayes, J.G. 1977, A new design
    formula for chain links. *NPL Report* NAC 79.

Barrodale, I. & Young, A. 1966, Algorithm for best L_1 and L_∞
    linear approximation on a discrete set. *Numer. Math.* **8**,
    295-306.

Gough, H.J., Cox, H.L. & Sopwith, D.G. 1934, Design of crane
    hooks and other components of lifting gear. *Proc. Inst.
    Mech. Eng.* **128**, 253-360.

3

Optimal control problems in tidal power generation

N. R. C. BIRKETT AND N. K. NICHOLS

1.   INTRODUCTION

The possibility of harnessing tidal energy has fascinated researchers for many years. Recently, tidal power generation schemes have attracted special attention due to international pressure for the development of 'alternative' energy sources. The problem of extracting power from tidal motion has been examined seriously in a number of countries, and in France a tidal power generating station has been successfully established at La Rance. In Great Britain investigation has concentrated on the production of energy from tidal flow in the Severn estuary, and a recent Government report (H.M.S.O. [1981]) has concluded that a barrage across the River Severn could be both technically and economically feasible. As expected with such a large engineering scheme, the arguments for and against the project have been hotly debated. To evaluate the tidal power scheme accurately it is necessary to calculate both the detailed component costs and the total attainable energy output, and assessment thus poses an extremely complicated optimization problem.

In this paper we develop global techniques which use the mathematical theory of control to investigate strategies for maximizing average power production from tidal schemes. In other studies, Berry [1982], Count [1980], H.M.S.O. [1981], Jefferys

53

[1981], Wilson *et al.* [1981], energy absorption figures for such schemes are computed primarily from simple linear models of one-dimensional flow in a rectangular basin. Wilson *et al.* [1981] examines the optimization of plant items, but does not include dynamic effects of flow in the basin in the analysis. Count [1980] derives optimal constant controllers for maximizing power output using a dynamic model with and without pumping, but in his scheme the control parameters are not allowed to vary with time. A describing function method is used by Jefferys [1981] and Berry [1982] to examine time-dependent controls with switches for a similar linear dynamic model, but in their work it is assumed that the flow parameters all vary harmonically with the tidal period at a single frequency. This technique is there-fore limited in not taking full account of all harmonic effects and is difficult to generalise to more sophisticated system models.

In the approach which we examine here, optimal control theory is applied to the full tidal power problem, and time-dependent strategies for maximizing energy output are determined using dynamic system models. This technique allows both the nature of the estuarine flow and fixed data for items of plant, such as turbines, sluices, barrier sites etc. to be taken into account while optimizing the engineering control parameters. To illustrate this approach we develop two models of the tidal power problem. Additional models of greater complexity and greater accuracy than these have also been constructed and treated by the same technique (see Birkett [1984], Birkett *et al.* [1984], Birkett & Nichols [1983]).

The first of the models was originally introduced by Dr B. Count for CEGB at the initial UCINA meeting in January, 1980, and was subsequently investigated by Mr N. Birkett in ful-fillment of the requirements for the Master of Science degree at Reading University, Birkett [1980]. This model does not represent tidal power schemes satisfactorily, but it does provide insight into the optimal control formulation of the problem and into appropriate solution methods. The second model more realistically describes controlled flow through a barrier but does not incorporate dynamic effects in the estuary. The system, in this case as in the first model, is represented by an ordinary differential equation, and the corresponding optimal control problems have a number of characteristics in common. In the more sophisticated models, dynamic effects in the estuary basin are taken into account and general estuary shapes with varying depth and cross-sectional area are treated. The resulting systems are represented by partial differential equations in these cases.

Numerical methods are required for the solution of the model problems, and we discuss here the development of appropriate computational techniques for each case. Two features are treated — the numerical solution of the differential system equations and the optimization of the control functions. For each model different techniques are necessary.

Investigation of the latter models has been carried out at Reading University under the support of an SERC CASE award with CEGB, and results are reported in full detail elsewhere, Birkett [1984], Birkett et al. [1984]. The feasibility of a global optimal control approach to the tidal power problem is established by this research, and further extensions of the techniques described here have been, and are continuing to be made to more realistic models, with the support of CEGB and SERC. Some preliminary results for the extended cases are given in Birkett & Nichols [1983].

2.   MODEL I — OSCILLATING SYSTEM

2.1   *Formulation of the Problem*

As a simple model of power generation, we consider the output obtained from a damped oscillating system. The system acts under an applied external force which can be controlled by switching the force on or off. The objective is to select the switching policy which maximizes the energy produced by the generator.

Mathematically the output $y(t)$ of the generator is given by the scalar second-order linear system equation

$$m\ddot{y} + (n + k)\dot{y} + by = g(t) \qquad (2\text{-}1)$$

where m is a mass associated with the system, n and k are natural damping constants associated with friction and the application of an electrical load, b is a constant associated with various 'spring forces' in the generating system, and $g(t)$ represents the external force providing the power. (All constants m, n, k and b are positive quantities and $b > (n + k)^2$ is assumed.) The energy E produced by the generator over time interval $[0, T]$ is given by

$$E = \int_0^T k\dot{y}^2 \, dt . \qquad (2\text{-}2)$$

The power generation problem is then defined as follows:

Given the initial state $(y(0), \dot{y}(0))$ of the system, determine the partition

$$\pi_N : 0 = t_0 < t_1 < \ldots < t_N = T$$

of the time interval $[0, T]$ which maximizes the output energy E, given by (2-2), over all π_N (any N), subject to the system equation (2-1) where

$$\begin{aligned} g(t) &= F_0 \sin \omega t , && \text{for} \quad t \in (t_j, t_{j+1}) , && j \in J \\ &= 0 , && \text{for} \quad t \in (t_j, t_{j+1}) , && j \notin J , \end{aligned} \qquad (2\text{-}3)$$

and J is any subset of the integers $\{0, 1, 2, \ldots, N-1\}$. The sinusoidal form of the forcing function is chosen to model the behaviour of a tidal flow force, but other forms for this function may be taken.

To apply optimal control theory we reformulate the problem by introducing a bounded scalar control function $u = u(t)$ which determines the proportion of the external force that is

applied at any moment. The differential equation (2-1) is then
scaled and rewritten as a first order system. For convenience
the system parameters are also simplified and we make the following definitions:

$$x_1(t) = y(t) \, , \quad x_2(t) = \dot{y}(t) \, , \quad K = n/m = k/m \, , \quad P^2 = b/m \, , \quad F_0/m = 1.$$

It is assumed that $P \geqslant K > 0$.

The optimal control problem then becomes

$$\max_{u} E(u) = \tfrac{1}{2} \int_0^T \boldsymbol{x}^T Q \boldsymbol{x} \; dt \qquad (2\text{-}4)$$

subject to

$$\dot{\boldsymbol{x}} = A\boldsymbol{x} + Bu \, , \qquad (2\text{-}5)$$

$$\boldsymbol{x}(0) = \boldsymbol{x}_0 \; \text{given} \, , \qquad (2\text{-}6)$$

where

$$\boldsymbol{x} = \left[x_1(t) \, , x_2(t) \right]^T$$

and

$$Q = \begin{bmatrix} 0 & 0 \\ 0 & 2K \end{bmatrix} , \quad A = \begin{bmatrix} 0 & 1 \\ -P^2 & -2K \end{bmatrix} , \quad B = \begin{bmatrix} 0 \\ f(t) \end{bmatrix} ,$$

with $f(t) = \sin \omega t$, $\omega < P$. Admissible controls are assumed to
belong to the set U_{ad} of measurable functions on $[0, T]$ satisfying (a.e.)

$$u(t) \in \Omega \equiv [0, 1] \, . \qquad (2\text{-}7)$$

We note that for any admissible control, the system equations
(2-5) − (2-6) have a unique continuous solution and that (2-4) −
(2-7) forms a classical constrained 'linear-quadratic' optimal
control problem.

The controlled proportion $u(t)$ of the applied external
force is allowed to vary over the entire closed, continuous interval $[0, 1]$. At first sight this problem is not strictly
equivalent to that initially described. In the next Section,
however, we derive necessary conditions for the solution which
show that the optimal control $u^*(t)$ maximizing the energy E
must be a 'bang-bang' control, that is

$$u^* = 1 \quad \text{for} \quad t \in (t_j, t_{j+1}), \quad j \in J$$
$$\quad = 0 \quad \text{for} \quad t \in (t_j, t_{j+1}), \quad j \notin J.$$

with some choice of J. The solution of the optimal control problem is therefore equivalent to that of the problem first formulated.

2.2   *Necessary Conditions for Optimality*

Necessary conditions for the solution of the optimal control problem (2-4)−(2-7) can be derived using either Pontryagin's Maximum Principle or a Lagrangian argument. We begin by applying the Maximum Principal in order to establish the 'bang-bang' nature of the optimal control. The following theorem provides the required result, Pontryagin *et al.* [1962] :

__Theorem.__ For control u with corresponding response $\boldsymbol{x}$, satisfying (2-5) and (2-6), to be *optimal*, it is necessary that there exists a continuous non-vanishing vector $(\lambda_0, \boldsymbol{\lambda}(t))$ with $\lambda_0 \geqslant 0$ such that if the Hamiltonian H is defined by

$$H(\boldsymbol{x}, u, \boldsymbol{\lambda}) = \lambda_0 \tfrac{1}{2} \boldsymbol{x}^T Q \boldsymbol{x} + \boldsymbol{\lambda}^T (A\boldsymbol{x} + Bu) \qquad (2\text{-}8)$$

then $\boldsymbol{\lambda}(t)$ satisfies

$$\dot{\boldsymbol{\lambda}} = -\partial H / \partial \boldsymbol{x} \equiv -\lambda_0 Q \boldsymbol{x} - A^T \boldsymbol{\lambda} \qquad (2\text{-}9)$$

$$\boldsymbol{\lambda}(T) = 0 \qquad (2\text{-}10)$$

and

$$H(\boldsymbol{x}, u, \boldsymbol{\lambda}) = \max_{v \in [0,1]} H(\boldsymbol{x}, v, \boldsymbol{\lambda}) \quad \text{(a.e.) on } [0, T].$$
$$\qquad (2\text{-}11)$$

We note that, given u, equations (2-5), (2-6) and (2-9), (2-10) together form a 4-dimensional linear two-point boundary value problem. The optimal control u must satisfy the Maximum Principle embodied in (2-11), that is, u must be such that

$$\boldsymbol{\lambda}^T Bu = \max_{v \in [0,1]} \boldsymbol{\lambda}^T Bv, \quad 0 \leqslant t \leqslant T, \qquad (2\text{-}12)$$

and, therefore, u takes the form

$$u = \begin{cases} 1 \\ 0 \end{cases} \quad \text{if} \quad \boldsymbol{\lambda}^T B \equiv \lambda_2 f(t) \begin{cases} > 0 \\ < 0 \,. \end{cases} \quad (2\text{-}13)$$

The optimal control thus lies on the boundary of the constraint set Ω of admissible values and is therefore a 'bang-bang' solution as required.

An exceptional case arises if $\boldsymbol{\lambda}^T B \equiv 0$ over any (measurable) subinterval of $[0, T]$, called a *singular arc*. On such a subinterval u is not defined by (2-11), and the optimal control may take interior values with $0 < u < 1$, as determined by the condition

$$\frac{d}{dt} (\boldsymbol{\lambda}^T B) \equiv 0 \,. \quad (2\text{-}14)$$

A further necessary condition, Bryson & Ho [1975], requires, however, that in such a case

$$-\frac{\partial}{\partial u} \left(\frac{d^2}{dt^2} (\boldsymbol{\lambda}^T B) \right) \leqslant 0 \quad (2\text{-}15)$$

must also hold for $E(u)$ to be maximized. For problem (2-4) — (2-7) it can be shown that, provided $f(t)$ is zero only at isolated points, then (2-15) cannot be satisfied along a singular arc. We conclude for this problem that the optimal solution contains no singular arcs, and is strictly 'bang-bang'.

Existence of an optimal solution to the control problem (2-4) — (2-7) can also be demonstrated for any continuously differentiable forcing function $f(t)$. Since the restraint set Ω is non-empty, convex and compact, and the system equations (2-5) are linear is $\boldsymbol{x}$ and u, the responses $\boldsymbol{x}(t)$ satisfying (2-5),(2-6) over all admissible controls $u(t) \in U_{ad}$ are uniformly bounded, that is

$$\max_{t \in [0, T]} |\boldsymbol{x}(t)| \leqslant \bar{X} < \infty \,, \quad (2\text{-}16)$$

where

$$|\boldsymbol{x}(t)| = \left[\sum_{1}^{n} x_i^2(t) \right]^{\frac{1}{2}} \,.$$

It follows then from Lee & Markus [1967] (Theorem 1, Chapt. 1) that there exists an optimal control belonging to U_{ad}.

To solve the optimal control problem $(2-4) - (2-7)$ we look for a piecewise constant control u such that the two-point boundary value problem $(2-5)$, $(2-6)$, $(2-9)$, $(2-10)$ together with $(2-13)$ is satisfied. Solutions are obtained by an iterative process described in the next Section. To establish the convergence of this process the gradient behaviour of the functional $E(u)$ is required, and it is therefore instructive to examine also the Lagrangian method of establishing necessary conditions for optimality.

The Lagrange functional associated with the problem $(2-4) - (2-7)$ is defined by

$$L(u) = \int_0^T \tfrac{1}{2} x^T Q x + \lambda^T (A x + B u - \dot{x}) \, dt \qquad (2-17)$$

where $\lambda(t)$ are Lagrange multipliers. For an admissible control u to be optimal it is necessary that the first variation $\delta L(u, v-u)$ of the functional L is *negative* for all admissible controls v, where δL is defined to be linear with respect to $\delta u = v - u$ and such that

$$L(v) - L(u) = \delta L(u, \delta u) + o(|v-u|) , \qquad (2-18)$$

(see Gelfand & Fomin [1963]). If we denote the difference between responses of the system $(2-5)$, $(2-6)$ to controls v and u by $\delta x(t)$, then taking variations and using integration by parts we find that

$$L(v) - L(u) = \int_0^T \lambda^T B \, \delta u + \tfrac{1}{2} \delta x^T Q \delta x \, dt , \qquad (2-19)$$

where λ satisfies the *adjoint* equations $(2-9)$, $(2-10)$, and thus

$$\delta L(u, \delta u) = \int_0^T \lambda^T B \, \delta u \, dt .$$

Assuming that $x(t)$ satisfies $(2-5)$, $(2-6)$, then the first variation of L with respect to u equals the first variation of E and we may write:

$$\delta L(u, \delta u) = \langle \nabla E(u), \delta u \rangle, \qquad (2\text{-}20)$$

where $\langle \cdot, \cdot \rangle$ is the inner product

$$\langle p, q \rangle = \int_0^T p(t)\, q(t)\, dt$$

and the function space gradient

$$\nabla E(u)(t) = \lambda^T(t) B.$$

For the control u to be optimal, then, it is necessary that

$$\langle \nabla E(u), v - u \rangle \equiv \int_0^T \lambda^T B(v-u)\, dt \leqslant 0 \qquad (2\text{-}21)$$

for all admissible controls v, and therefore u must take the piecewise constant form (2-13), as previously derived (see also Lions [1971]).

2.3   *Solution of the Optimal Control Problem*

To solve the optimal control problem $(2\text{-}4) - (2\text{-}7)$ we use an iterative technique for determining a piecewise constant admissible control u of form (2-13) with corresponding response x and adjoint λ satisfying the system and adjoint equations (2-5), (2-6) and (2-9), (2-10). The iteration is described by the following:

Algorithm 1

<u>Step 1.</u>   Choose $u^0 \in U_{ad}$, piecewise constant, such that

$$u^0(t) := 0 \text{ or } 1, \quad \forall t \in [0, T]$$

Set $E^{-1} := 0$.

<u>Step 2.</u>   *For* $k = 0, 1, 2, \ldots$ *do*

<u>Step 2.1</u>   Solve
   (i) $\dot{x}^k = Ax^k + Bu^k, \quad x^k(0) = x_0$,
   (ii) $\dot{\lambda}^k = -A^T \lambda^k - Qx^k, \quad \lambda^k(T) = 0$.

<u>Step 2.2</u>   Evaluate

$$E^k := \int_0^T \tfrac{1}{2} x^{k^T} Q x^k \, dt.$$

<u>Step 2.3</u> If $E^k - E^{k-1} <$ tol *then go to Step 3.*

<u>Step 2.4</u> Set

$$u^{k+1} := \begin{cases} 1 & \text{if } \lambda^T B \geqslant 0 \\ 0 & \text{otherwise .} \end{cases}$$

<u>Step 2.5</u> CONTINUE

<u>Step 3.</u> Set $u := u^k$ and STOP.

The algorithm generates a sequence of admissible controls $\{u^k\}$ for which the functionals $E^k \equiv E(u^k)$ are monotonically non-decreasing. It is easily seen from (2-18) that if u^{k+1} is given by *Step 2.4* and x^{k+1} is the corresponding response, then

$$E^{k+1} - E^k = \int_0^T \lambda^{k^T} B \left(u^{k+1} - u^k \right)$$
$$+ \tfrac{1}{2} \left(x^{k+1} - x^k \right)^T Q \left(x^{k+1} - x^k \right) dt \geqslant 0,$$

$$(2\text{-}22)$$

since Q is positive and semi-definite. The sequence $\{E^k\}$ is also bounded, since the responses x^k satisfy (2-16), and there-fore, there exists E^* such that

$$\lim_{k \to \infty} E(u^k) = E^*.$$

Furthermore there exists a subsequence of $\{u^k\}$ which converges weakly to $u^* \in U_{ad}$, (since U_{ad} is weakly compact (sequentially), Lee & Markus [1967]), with responses which converge pointwise to x^*, the response to u^*. It follows that $E^* = E(u^*)$, and the iteration process described by the algorithm is convergent. By the definition of u^{k+1} we also have

$$\sup_{u \in U_{ad}} \left\langle \nabla E(u^k), u - u^k \right\rangle \leqslant \int_0^T \lambda^{k^T} B \left(u^{k+1} - u^k \right) dt \leqslant E^{k+1} - E^k,$$

$$(2\text{-}23)$$

and by the convergence of $\{E^k\}$ it can then be shown that

$$\sup_{u \in U_{ad}} \left\langle \nabla E(u^*), u - u^* \right\rangle \leqslant 0, \qquad (2\text{-}24)$$

and, therefore, the limiting control u^* satisfies the necessary conditions for optimality, Birkett [1984]. We note that the algorithm could also be stopped when

$$\left\langle \nabla E\left(u^{k}\right),\ u^{k+1} - u^{k}\right\rangle \equiv \int_{0}^{T} \lambda^{k^{T}} B\left(u^{k+1} - u^{k}\right) dt < \text{tol},$$

$$(2\text{-}25)$$

that is, when u^{k} is close to satisfying the necessary conditions for optimality.

We observe that Algorithm 1 defines a function iteration, and in practice it is necessary to discretize the procedure in order to compute the iterates numerically. The interval $[0,T]$ is partitioned into N steps of length $h = T/N$, and solutions are determined at the mesh points $t_{j} = jh$. A finite difference technique is used in *Step 2.1* to solve the state and adjoint systems, and a quadrature rule is used in *Step 2.2* to evaluate the functional E^{k}. The differential equations are approximated on the mesh by the trapezoidal difference scheme and the functional is approximated by the trapezium rule. The state equations are integrated forward from $x(0) = x_{0}$, and the adjoint equations are then integrated backward from $\lambda(T) = 0$. Since the system equations are linear in x and λ, the trapezoidal scheme can be written as a one-step 'explicit' method, and, since meshes with constant step-size are used, no interpolations between mesh point values are required. The discrete equations for *Step 2.1* and *Step 2.2* are thus given by

Step 2.1'

(i) $\quad x_{0}^{k} := x(0)$,

$$x_{j+1}^{k} := \left(I - \tfrac{1}{2}hA\right)^{-1}\left[\left(I + \tfrac{1}{2}hA\right)x_{j}^{k} + \tfrac{1}{2}h\left(B_{j}u_{j}^{k} + B_{j+1}u_{j+1}^{k}\right)\right],$$

$$j = 0, 1, \ldots, N-1;$$

(ii) $\quad \lambda_{N}^{k} := 0$,

$$\lambda_{j-1}^{k} := \left(I - \tfrac{1}{2}hA^{T}\right)^{-1}\left[\left(I + \tfrac{1}{2}hA^{T}\right)\lambda_{j}^{k} - \tfrac{1}{2}\left(Qx_{j}^{k} + Qx_{j-1}^{k}\right)\right],$$

$$j = N, N-1, \ldots, 1;$$

Step 2.2'

$$E^{k+1} := \tfrac{1}{2}h \sum_{j=0}^{N}{}'' \ x_{j}^{k^{T}} Q x_{j}^{k} ,$$

where
$$u_j^k \cong u^k(t_j), \quad \boldsymbol{x}_j^k \cong \boldsymbol{x}^k(t_j), \quad \lambda_j^k \cong \lambda^k(t_j) \quad \text{and} \quad B_j = B(t_j).$$
The discrete values of the control are determined in *Step 2.4* at each point t_j of the mesh by

$$u_j^{k+1} = 1 \quad \text{if} \quad \lambda_j^{k^T} B_j \geqslant 0, \quad \text{and} \quad u_j^{k+1} = 0 \quad \text{otherwise.}$$

Both numerical integrations (i) and (ii) are absolutely stable since the eigenvalues μ_i of A are just

$$\mu_{1,2} = -K \pm i\sqrt{P^2 - K^2},$$

and therefore $\mathrm{Re}(\mu_i) < 0$, which is sufficient for absolute stability, Lambert [1973]. The scheme is only second-order, but since very little storage is required, and the computations are simple, it is feasible to take very small step-sizes in order to achieve high accuracy.

2.4 *Numerical Results*

The numerical procedures were tested for various choices of the parameters P, K and also for different choices of the forcing function $f(t)$. In most calculations the value of T was taken to be π, and the initial approximation was chosen to be $u^0 = 1$ on $[0, T]$, so that the result from the first iteration corresponded to the *uncontrolled* system. Experiments were also performed with initial approximations containing up to 100 switch points, in order to determine whether the converged results were affected by the starting choice.

In all cases tested the numerical value of the functional $E(u)$ increased at every iteration until consistent values were obtained, and the switch points, which essentially determine the solution, were also seen to converge. These results are illustrated in Table 1 for the case $P = 2$, $K = 1$, $\boldsymbol{x}_0 = \boldsymbol{0}$, $f(t) = \sin t$, $h = \pi/800$. Solutions in all cases were computed for a decreasing sequence of mesh sizes h. Convergence was observed and consistency was achieved for sufficiently small h (with a sufficient number of iterations for convergence of the functional). Table 2 shows the converged values of E for various meshes with N steps, together with the number of iterations required for convergence.

TABLE 1
Convergence of Iteration
$P = 2$, $K = 1$, $h = \pi/800$

Iteration no.	$E(u^k)$	Switch points	
		t_1	t_2
$k = 0$	0.593	0	0
1	0.1011	0.0490	2.2044
2	0.1035	0.5541	2.1306
3	0.1039	0.5751	2.1003
.	.	.	.
.	.	.	.
8	0.1041	0.5768	2.0711
9	0.1041	0.5746	2.0707
10	0.1041	0.5746	2.0707

TABLE 2
Convergence of Discretization
$P = 2$, $K = 1$

No. of Mesh Steps	$E(u^*)$	No. of Iterations
100	0.1025	6
200	0.1034	6
400	0.1038	8
800	0.1041	10
1600	0.1042	12

It was also found that the initial choice u did not affect the final result of the iterations and that the qualitative behaviour of the numerical solutions was as expected. In Figures 1 and 2 solution curves are shown for different values of P and K with $f(t) = \sin t$. In the first case the optimal control contains only two switch points, but in the second case, where $P \gg K$, the optimal control switches on and off at approximately the natural frequency of the state system $(P/2\pi$ cycles per unit time). Similar behaviour is seen in Figure 3, which shows solution curves for the same system with forcing function

$$f(t) = \sin t + \tfrac{1}{2} \sin 2t + \sin 3t .$$

This type of behaviour agrees with the intuitive expectation that the control should exploit resonance effects in order to obtain maximum energy.

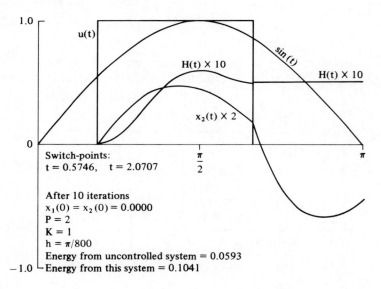

Figure 1   MODEL 1

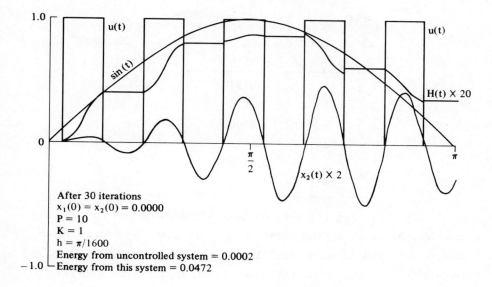

Figure 2 MODEL I

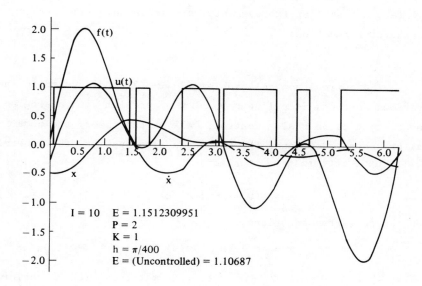

Figure 3 MODEL I — $f(t) = \sin t + \frac{1}{2} \sin 2t + \sin 3t$

Table 3 *Comparison of Energy Production*

P	K	E uncontrolled	E optimal
2	1	0.0593	0.1042
10	1	0.0002	0.0472
20	2	0.00002	0.0330
2	2	0.0515	0.0750
4	4	0.0132	0.0269
10	10	0.0013	0.0036

The most significant conclusion regarding power generation that can be reached from these results is that the energy which can be extracted from an oscillating system of type (2-1) can be more than doubled with the use of an appropriate control strategy. In Table 3 the energy from the uncontrolled and optimally controlled systems for various P, K and $f(t) = \sin t$ are shown. For the simple harmonic forcing function large improvements can be made in the maximum average energy produced; similar, but smaller improvements, are achievable for less regular forcing functions.

2.5   *Generalization*

The optimal control problem (2-4) − (2-7) discussed in the previous sections is a linear quadratic problem with a single input, that is, a scalar control. All of the theory and numerical procedures developed for this case also apply to the general *multi*-input optimization problem

$$\max_{\pmb{u}} E(\pmb{u}) = \frac{1}{2} \int_0^T \pmb{x}^T Q \pmb{x}\, dt, \qquad (2\text{-}4')$$

subject to system equations

$$\dot{\pmb{x}} = A\pmb{x} + B\pmb{u}, \quad \pmb{x}(0) = \pmb{x}_0, \qquad (2\text{-}5'),(2\text{-}6')$$

where $\pmb{x}$ is an $\tilde{n}$-vector, A,Q are $\tilde{n} \times \tilde{n}$ matrices with Q positive semi-definite, $B = B(t)$ is an $\tilde{n} \times \tilde{m}$ continuously differentiable matrix function of full rank for all t, and $\boldsymbol{u}$ is an $\tilde{m}$-dimensional measurable control vector $(\tilde{m} \leqslant \tilde{n})$ on $[0,T]$ such that (a.e.)

$$0 \leqslant u_{i}(t) \leqslant 1 , \quad i = 1,2,\ldots,\tilde{m} . \qquad (2\text{-}7')$$

The necessary conditions imply that the optimal control is component-wise 'bang-bang' in nature, and the switches occur where the corresponding components of $\nabla E(\boldsymbol{u})$ change sign. (There are no singular arcs, provided $B^{T}QB$ is singular only at isolated points.) The techniques developed are applicable, therefore, to a wide range of general problems.

3.   MODEL II — FLAT BASIN

3.1   *Formulation of the Control Problem*

The oscillating system model described in Section 2 provides insight into the formulation and solution of optimal control problems in power generation, and shows that energy extraction can be greatly enhanced by the use of an appropriate control strategy. This model does not, however, represent satisfactorily a tidal power generating system in which energy is extracted from flow across a barrier.

To model the tidal power problem we start with a simple one-dimensional system consisting of an estuary with a barrier at the origin, and a basin upstream of the barrier, as shown in Figure 4. The surface elevation above mean height is assumed constant throughout the basin at any point in time; that is, dynamic effects in the basin are ignored and the basin surface is assumed to remain flat. Flow is permitted across the barrier, through turbines only, and the influx velocity is assumed proportional to the head difference between the tidal elevation on the seaward side of the barrier and the surface elevation in the basin. The control determines the proportional discharge across the available turbines.

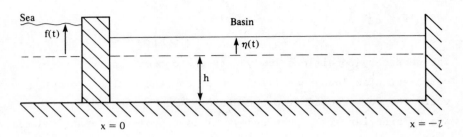

Figure 4 Tidal Basin

Mathematically, then, the influx velocity u_0 is given by

$$u_0 = \alpha(t) \left[f(t) - \eta(t) \right] , \qquad (3\text{-}1)$$

where $\alpha(t)$ is the influx control, which is bounded such that
$0 \leqslant \alpha \leqslant a_0$ for all t, $\eta(t)$ is the basin elevation above mean
and $f(t)$ is the tidal elevation above mean. By the law of
conservation of mass we must also have

$$u_0 = l\dot{\eta}(t)/h , \qquad (3\text{-}2)$$

where l is the length of the basin and h is the mean depth in
the basin. Solutions of the system equations $(3\text{-}1) - (3\text{-}2)$ may
be determined over any interval $[0, T]$, starting from any initial
state of the system. For the tidal problem we are specifically
interested in a tidal period, however, where the forcing $f(t)$ is
a dominant harmonic of form

$$f(t) = F_0 \cos \omega t , \quad \omega = 2\pi/T , \qquad (3\text{-}3)$$

with period $T \simeq 12 \times 3600$ s. We therefore expect to find 'steady-
state' solutions to the system equations which are also periodic
on $[0, T]$, and we impose the boundary conditions $\eta(0) = \eta(T)$ on
the problem.

The instantaneous power output from the turbine is
assumed proportional to the flow times the head difference, that

is, $u_0[f(t) - \eta(t)]$, and the average power generated over time
interval $[0,T]$ is therefore given by

$$\bar{P} = \rho g \frac{S}{T} \int_0^T \alpha(t)[f(t) - \eta(t)]^2 \, dt, \qquad (3\text{-}4)$$

where S is the surface area of the barrier, g is gravitational
acceleration, and ρ is the fluid density.

The optimal control problem is then to determine the control $\alpha(t)$ which maximizes the average power $\bar{P}$ subject to the
system equations and boundary conditions. The equations may be
normalized with respect to time, and the optimization problem
is then given by

$$\max_{\alpha \in U_{ad}} E(\alpha) = \int_0^1 \alpha(t)[f(t) - \eta(t)]^2 \, dt, \qquad (3\text{-}5)$$

subject to

$$\dot{\eta} = K\alpha(t)[f - \eta], \quad K = hT/l . \qquad (3\text{-}6)$$

$$\eta(0) = \eta(1) , \qquad (3\text{-}7)$$

where U_{ad} is the set of measurable functions $\alpha(t)$ on $[0,1]$
such that (a.e.)

$$\alpha(t) \in \Omega \equiv [0, a_0] . \qquad (3\text{-}8)$$

3.2   *Conditions for the Optimal*

For the control problem $(3\text{-}5) - (3\text{-}8)$ to be well posed,
it is necessary that, for any given admissible control function
$\alpha(t) \in U_{ad}$, the system equation $(3\text{-}6)$ together with boundary
condition $(3\text{-}7)$ must have a unique solution, continuously dependent on the data of the problem. If this can be shown, then
necessary conditions for the optimal can be derived using the
Maximum Principle or the Lagrange technique, and the existence
of an optimal control $\alpha^*(t) \in U_{ad}$ can be established by the
same arguments as used for the oscillating system in Section 2.

The system equation $(3\text{-}6)$ is linear in the state variable
$\eta(t)$ and it is easily demonstrated, therefore, that for any

$$\alpha(t) \in U_{ad} \quad \text{with} \quad \int_0^1 \alpha(s)\,ds > 0,$$

that is $\alpha(t) \neq 0$, the unique solution of (3-6) is given by

$$\eta(t) = \frac{\phi(t)}{1 - \phi(1)} \int_0^1 \phi(1)\,\phi^{-1}(s)\,K\alpha(s)\,f(s)\,ds$$

$$+ \phi(t) \int_0^t \phi^{-1}(s)\,K\alpha(s)\,f(s)\,ds \qquad (3-9)$$

where

$$\phi(t) = \exp\left(-\int_0^t K\alpha(s)\,ds\right).$$

The case $\alpha \equiv 0$ is trivial in the sense that no fluid crosses the barrier and no power is generated, and it may thus be disregarded. From the expression (3-9), it follows that if the forcing function $f(t)$ is continuously differentiable and bounded on $[0,1]$, then the response $\eta(t)$ of the system (3-6) is uniformly bounded over all non-trivial $\alpha \in U_{ad}$ (see Birkett [1984]), and a (non-trivial) optimal control belonging to the set of admissible controls exists.

From the Maximum Principle we find that a necessary condition for an admissible control $\alpha(t)$ and its response $\eta(t)$ to be optimal is the existence of an adjoint $\lambda(t)$ satisfying

$$\dot{\lambda} = K\alpha(t)\lambda + 2\alpha(t)[f(t) - \eta(t)] \qquad (3-10)$$

$$\lambda(0) = \lambda(1) \qquad (3-11)$$

and such that

$$H = \alpha[f - \eta]^2 + \lambda K\alpha[f - \eta]$$

is maximized with respect to $\alpha \in \Omega$ for $t \in [0,1]$. Hence the optimal is 'bang-bang' in nature and is given by

$$\alpha(t) = \begin{cases} a_0 & \text{if } [f - \eta]^2 + K\lambda[f - \eta] \geqslant 0 \\ 0 & \text{otherwise} \end{cases} \qquad (3-12)$$

The Lagrangian for problem (3-5) − (3-6) is defined by

$$L(\alpha) = \int_0^1 \left(\alpha[f - \eta]^2 + \lambda\,(K\alpha[f - \eta] - \dot{\eta})\right) dt \qquad (3-13)$$

and, taking variations, we find that if $\eta(t)$ and $\lambda(t)$ satisfy the system and adjoint equations (3-6), (3-7), (3-10), and (3-11), then

$$L(\beta) - L(\alpha) = \int_0^1 \Big(([f-\eta]^2 + K\lambda[f-\eta])\, \delta\alpha$$
$$- (2[f-\eta] + K\lambda)\, \delta\alpha\, \delta\eta + \alpha(\delta\eta)^2 + \delta\alpha(\delta\eta)^2 \Big) dt$$

(3-14)

where β is any non-trivial admissible control, $\delta\alpha = \beta - \alpha$ and $\delta\eta$ is the difference between the responses to controls β and α. The first variation is therefore given by

$$\delta L \equiv \langle \nabla E(\alpha),\, \beta - \alpha \rangle = \int_0^1 ([f-\eta]^2 + K\lambda[f-\eta])(\beta-\alpha)\, dt \qquad (3-15)$$

which is non-positive for all $\beta \in U_{ad}$ only if $\alpha(t)$ satisfies (3-12). The optimal control must, thus, be piecewise constant with values at the extremes of the constraint set Ω and switches at zeros of the function space gradient

$$\nabla E(\alpha) = [f-\eta]^2 + K\lambda[f-\eta]. \qquad (3-16)$$

We note that provided the derivative of the forcing function, $\dot{f}(t)$, has only isolated zeros, the optimal solution can contain no singular arcs, since $\nabla E(\alpha) \equiv 0$ over some sub-interval of $[0,T]$ implies that either $\dot{f} \equiv 0$, or $f \equiv \eta$ and therefore, from (3-6) $\dot{\eta} \equiv 0 \equiv \dot{f}$, on that sub-interval, which contradicts the assumption on $\dot{f}$. The existence and general 'bang-bang' nature of the optimal control is thus established.

3.3 *Numerical Solution of the Control Problem*

To determine an admissible control $\alpha(t)$ with corresponding response $\eta(t)$ and adjoint $\lambda(t)$ satisfying the necessary conditions (3-6), (3-7), and (3-10) — (3-12), we again use an iterative technique similar to that defined by Algorithm 1. If we have an approximation α^k to the optimal control, with response and adjoint η^k, λ^k satisfying the system and adjoint equations (3-6), (3-7) and (3-10), (3-11), then we may choose a new approximation

$$\tilde{\alpha}^{k+1} : = a_0 \quad \text{if} \quad \nabla E(\alpha^k) \geqslant 0 \tag{3-17}$$

$$: = 0 \quad \text{otherwise,}$$

where $\nabla E(\alpha)$ is given by (3-16). This selection maximizes the first variation $\langle \nabla E(\alpha^k), \tilde{\alpha}^{k+1} - \alpha^k \rangle$ of the functional (3-5) over all possible choices for $\tilde{\alpha}^{k+1}$. The energy functional (3-5) is not now quadratic, however, and it can be seen from (3-14) that $E(\tilde{\alpha}^{k+1}) - E(\alpha^k) \equiv L(\tilde{\alpha}^{k+1}) - L(\alpha^k)$ is not necessarily non-negative for this choice of $\tilde{\alpha}^{k+1}$. It can be shown, however, that there does exist an admissible control $\alpha^{k+1} = \alpha^k + \theta(\tilde{\alpha}^{k+1} - \alpha^k)$ for some $\theta \in [0,1]$ such that $E(\alpha^{k+1}) \geqslant E(\alpha^k)$ (see Gruver & Sachs [1980]). We can therefore construct a sequence of controls $\{\alpha^k\}$ for which the functionals $E^k = E(\alpha^k)$ are monotonically non-decreasing. Since the responses η^k are continuously dependent on the controls α^k and are uniformly bounded for all non-trivial $\alpha \in U_{ad}$, we can again show that the sequence $\{E^k\}$ is bounded and convergent. As in the case of the oscillating system, there also exists a subsequence of the controls which converges weakly to a (non-trivial) limit $\alpha^* \in U_{ad}$ such that

$$\lim_{k \to \infty} E^k = E(\alpha^*) , \tag{3-18}$$

and

$$\sup_{\alpha \in U_{ad}} \langle \nabla E(\alpha^*), \alpha - \alpha^* \rangle \leqslant 0 , \tag{3-19}$$

for all admissible controls α, and α^* satisfies the necessary conditions for the optimal.

In order to obtain periodic solutions to the state and adjoint equations (3-6) and (3-10) for each α^k, inner iteration procedures are used. We define the processes

$$\eta^{m+1}(0) = \eta^m(1) \equiv F(\eta^m(0)) , \quad m = 0,1,2,\ldots , \tag{3-20}$$

$$\lambda^{r+1}(1) = \lambda^r(0) \equiv G(\lambda^r(1)) , \quad r = 0,1,2,\ldots , \tag{3-21}$$

where $F(\eta(0))$ and $G(\lambda(1))$ are the solutions of the differential equations (3-6) and (3-10) with given initial data $\eta(0)$ at $t=0$ and $\lambda(1)$ at $t=1$, respectively. The operators F, G can be written

$$F(\eta(0)) = \phi(1)\,\eta(0) + \int_0^1 \phi(1)\,\phi^{-1}(s)\,\alpha(s)\,f(s)\,ds\,, \qquad (3\text{-}22)$$

$$G(\lambda(1)) = \psi(0)\,\lambda(1) + \int_0^1 \psi(0)\,\psi^{-1}(s)\,2\alpha(s)\,(f(s)-\eta(s))\,ds\,, \qquad (3\text{-}23)$$

where

$$\phi(t) = \exp\left(-\int_0^t K\alpha(s)\,ds\right) \quad \text{and} \quad \psi(t) = \exp\left(-\int_t^1 K\alpha(s)\,ds\right),$$

and it can be seen that since

$$|F(v)-F(w)| \leqslant \phi(1)\,|v-w|\,, \quad 0 < \phi(1) < 1\,, \qquad (3\text{-}24)$$

$$|G(v)-G(w)| \leqslant \psi(0)\,|v-w|\,, \quad 0 < \psi(0) < 1\,, \qquad (3\text{-}25)$$

the operators F and G are both contractions (see Keller [1968]). The iterations (3-20) and (3-21), therefore, both converge to the unique fixed points η_0, λ_1 of the operators, and the solutions of the equations (3-6), (3-10) with initial data $\eta(0) = \eta_0$, $\lambda(1) = \lambda_1$ thus also satisfy the periodic boundary conditions

$$\eta(1) = F(\eta_0) = \eta_0 = \eta(0)\,, \quad \lambda(0) = G(\lambda_1) = \lambda_1 = \lambda(1)\,. \qquad (3\text{-}26)$$

The complete function iteration procedure for determining the optimal control is described by the following:

Algorithm 2

Step 1. Choose $\alpha^0 \in U_{ad}$, piecewise constant such that

$$\alpha^0 := 0 \quad \text{or} \quad a_0, \quad \forall t \in [0,1]\,.$$

Choose $\eta^0(0)$, $\lambda^0(1)$.

Set $\quad \theta := 1$, $E^0 := 0$, $\tilde\alpha^1 := \alpha^0$.

Step 2. For $k := 1, 2, \ldots,$ do

Step 2.1 Set $\alpha^k := \alpha^{k-1} + \theta(\tilde\alpha^k - \alpha^{k-1})$.

Step 2.2 Set $\tilde\eta^0(1) := \eta^{k-1}(0)$, $\tilde\lambda^0(0) := \lambda^{k-1}(1)$.

<u>Step 2.3</u> *For* $m: = 1, 2, \ldots, do$

   <u>Step 2.3.1</u> Solve $\dot{\tilde{\eta}}^m = K\alpha^k[f - \tilde{\eta}^m]$, $\quad \tilde{\eta}^m(0) = \tilde{\eta}^{m-1}(1)$.

   <u>Step 2.3.2</u> *If* $|\tilde{\eta}^m(1) - \tilde{\eta}^{m-1}(1)| < $ tol *then* set $\eta^k \equiv \tilde{\eta}^m$ *and goto Step 2.4* .

   <u>Step 2.3.3</u> CONTINUE.

<u>Step 2.4</u> *For* $m: = 1, 2, \ldots, do$

   <u>Step 2.4.1</u> Solve $\dot{\tilde{\lambda}}^m = K\alpha^k \tilde{\lambda}^m + 2\alpha^k[f - \eta^k]$,
$$\tilde{\lambda}^m(1) = \tilde{\lambda}^{m-1}(0) .$$

   <u>Step 2.4.2</u> *If* $|\tilde{\lambda}^m(0) - \tilde{\lambda}^{m-1}(0)| < $ tol
then set $\lambda^k \equiv \tilde{\lambda}^m$ *and goto Step 2.5*.

   <u>Step 2.4.3</u> CONTINUE

<u>Step 2.5</u> Evaluate
$$\nabla E^k: = [f - \eta^k]^2 + K\lambda^k[f - \eta^k] ,$$
$$E^k: = \int_0^1 \alpha^k[f - \eta^k]^2 dt .$$

<u>Step 2.6</u> Set
$$\tilde{\alpha}^{k+1}: = \begin{cases} 1 & \text{if } \nabla E^k \geqslant 1 \\ 0 & \text{otherwise} . \end{cases}$$

<u>Step 2.7</u> *If* $\langle \nabla E^k, \tilde{\alpha}^{k+1} - \alpha^k \rangle < $ tol *then goto Step 3* .

<u>Step 2.8</u> *If* $E^k \leqslant E^{k-1}$ *then* set $\theta: = \theta/2$
and goto Step 2.1.

<u>Step 2.9</u> CONTINUE .

<u>Step 3.</u> Set $\alpha: = \alpha^k$ *and* STOP.

In practice the function iteration described by Algorithm 2 is replaced by a discretized process. The interval $[0,1]$ is divided into steps of length $h = 1/N$ and solutions are determined at mesh points $t_j = jh$. The initial value problems in *Step 2.3.1* and *Step 2.4.1* are solved by a finite difference method using the trapezoidal scheme, and the functionals E^k and

$\langle \nabla_E{}^k , \; \tilde{\alpha}^{k+1} - \alpha^k \rangle$ are approximated in *Step 2.5* and *Step 2.7* by
the trapezium quadrature rule. The difference approximations
are given by

$$\tilde{\eta}_0^m := \tilde{\eta}_N^{m-1} \; , \tag{3-27}$$

$$\tilde{\eta}_{j+1}^m := \left(1 + \tfrac{1}{2} Kh\alpha_{j+1}^k\right)^{-1} \left[\left(1 - \tfrac{1}{2} Kh\alpha_j^k\right) \tilde{\eta}_j^m \right.$$
$$\left. + \tfrac{1}{2} Kh\left(\alpha_{j+1}^k f_{j+1} + \alpha_j^k f_j\right) \right] \; ,$$
$$j = 0,1,\ldots,N-1$$

and

$$\tilde{\lambda}_N^m := \tilde{\lambda}_0^{m-1} \tag{3.28}$$

$$\tilde{\lambda}_{j-1}^m := \left(1 + \tfrac{1}{2} Kh\alpha_{j-1}^k\right)^{-1} \left[\left(1 - \tfrac{1}{2} Kh\alpha_j^k\right) \tilde{\lambda}_j^m \right.$$
$$\left. - Kh\left(\alpha_{j-1}^k \left(f_{j-1} - \eta_{j-1}^k\right) + \alpha_j^k\left(f_j - \eta_j^k\right)\right) \right] \; ,$$
$$j = N,\; N-1,\ldots,1$$

and the quadrature rule gives

$$E^k := h \sum_{j=0}^{N}{}'' \alpha_j^k \, [f_j - \eta_j^k]^2 \; , \tag{3-29}$$

and

$$\langle \nabla_E{}^k , \; \tilde{\alpha}^{k+1} - \alpha^k \rangle := h \sum_{j=0}^{N}{}'' \nabla_{E_j}^k \left(\alpha_j^{k+1} - \alpha_j^k\right) \; , \tag{3-30}$$

where $\alpha_j^k, \tilde{\alpha}_j^k, \eta_j^k, \lambda_j^k, \nabla_{E_j}^k$ are approximations to the values of
$\alpha^k,\; \tilde{\alpha}^k,\; \eta^k,\; \lambda^k$ and $\nabla_E{}^k$ at the mesh point t_j, and $f_j = f(t_j)$.
The discrete values α_j^k and $\tilde{\alpha}_j^k$ are determined in a natural way
in *Step 2.1* and *Step 2.6* .

The difference schemes (3-27) and (3-28) are both absolutely stable and the iterations of *Step 2.3* and *Step 2.4* both
converge to the solution of discrete periodic boundary value
problems which approximate the state and adjoint systems (3-6),
(3-7) and (3-10), (3-11). The difference schemes are technically
second-order, but since the first derivative of the state variable η generally contains discontinuities, we cannot expect

second-order asymptotic behaviour in this case. It can be shown,
however, that the schemes for this case are at least first-order
accurate, Birkett [1984], and therefore are convergent as $h \to 0$.

3.4   *Results*

Solutions to the optimization problem (3-5) − (3-8) were
computed using the numerical procedures discussed in Section 3.3
with data typical of the Severn Estuary. The parameter values
are given by

$$T = 4.32 \times 10^4 \text{ s.}$$
$$l = 5.0 \ \times 10^4 \text{ s.}$$
$$h = 15 \text{ m.}$$
$$S = 22.5 \times 10^4 \text{ m}^2$$

$$(3-31)$$

With this data $K = 12.96$ for the normalized problem, and the
imposed tidal elevation is given by $f(t) = \cos 2\pi t$. The maximum
discharge across the barrier depends on a_0 , which is typically
in the range $0.1 \leqslant a_0 \leqslant 1$. The average power in watts is determined from

$$\bar{P} = \rho g S F_0^2 E(\alpha) \qquad\qquad (3-32)$$

Solutions were also obtained for other representative
values of the parameters K and a_0 in order to examine the behaviour of the numerical procedures. In all tests it was found
that Algorithm 2 was convergent and required less than twenty
iterations to reduce the first order correction to the functional
E to 1% of its value. Frequently the choice $\theta = 1$ was sufficient throughout the iteration. The solutions were also observed
to converge as the discretization step $h \to 0$, in all examples.
Typical solution curves are shown in Figure 5 for the data
(3-31) with $a_0 = 1$ and $h = 1/400$.

The optimal *constant* control for the problem (3-5) −
(3-8) is derived by Count [1980] as $\alpha(t) \equiv 2\pi/K$ in the case
$a_0 = 1$. This choice of $\alpha(t)$ was taken as the initial approximation for the iteration procedure, and the average power output

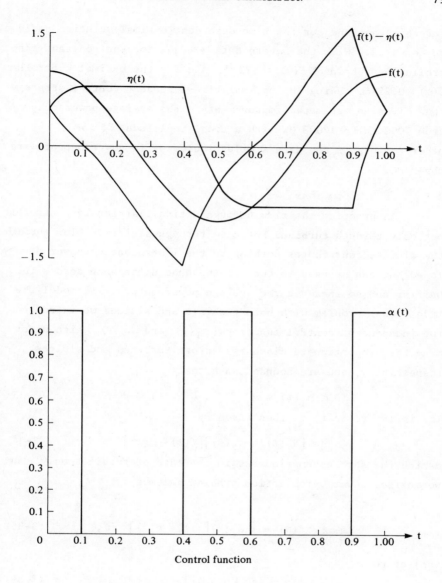

Figure 5 MODEL II — Turbines only

for this case and for the time-dependent optimal solution could
be compared. With the Severn data (3-31), for the constant controller we obtained $E(\alpha) = 0.1212$, and for the optimal controller
$E(\alpha) = 0.2276$. Using the optimal time-dependent control strategy
thus gives an estimated improvement in the average power output
from 7000 MW to 12000 MW with a tidal amplitude $F_0 = 5\text{m}$. For
other choices of the data similar increases in power output were
observed.

3.5   Generalizations

In practice the flow through a tidal barrier is controlled
not only through turbines but also through sluices. Flow through
the sluices contributes nothing to the instantaneous power developed but can be used to control the head difference across the
barrier and so increase the average power output. To model the
tidal power problem with both turbines and sluices we introduce
two independent control functions $\alpha_1(t)$ and $\alpha_2(t)$ which determine the proportionate discharge across turbines and sluices,
respectively, and are bounded such that

$$0 \leqslant \alpha_1(t) \leqslant a_1, \qquad 0 \leqslant \alpha_2(t) \leqslant a_2. \tag{3-33}$$

The influx velocity is then given by

$$u_0 = [\alpha_1(t) + \alpha_2(t)][f(t) - \eta(t)]. \tag{3-34}$$

Making the same assumptions as in the case of turbines only, the
normalized power optimization problem becomes

$$\max_{\alpha \in U_{ad}} E(\alpha) = \int_0^1 \alpha_1(t)[f(t) - \eta(t)]^2 \, dt, \tag{3-35}$$

subject to

$$\dot{\eta} = (\alpha_1 + \alpha_2) K [f - \eta], \tag{3-36}$$

$$\eta(0) = \eta(1), \tag{3-37}$$

where U_{ad} is now the set of measurable two-dimensional vector
functions $\alpha = [\alpha_1, \alpha_2]^T$ satisfying (3-33) (a.e.)

The results of Section 3.2 are easily extended to the

problem $(3-35) - (3-37)$, and it can be shown that for any non-trivial $\alpha \in U_{ad}$, there is a unique solution satisfying the system equations $(3-36)$, $(3-37)$, that the responses η to $(3-36)$, $(3-37)$ are uniformly bounded for all non-trivial $\alpha \in U_{ad}$, and that an optimal control $\alpha^* \in U_{ad}$ exists. The gradient function $\nabla E(\alpha)$ is now given by

$$\nabla E(\alpha) = \left[(f-\eta)^2 + \lambda K(f-\eta), \quad \lambda K(f-\eta) \right]^T \qquad (3-38)$$

where the response satisfies the system equations $(3-36)$, $(3-37)$ and the adjoint λ satisfies

$$\dot{\lambda} = \left[\alpha_1 + \alpha_2\right] K\lambda + 2\alpha_1\left[f-\eta\right], \qquad (3-39)$$

$$\lambda(0) = \lambda(1). \qquad (3-40)$$

Necessary conditions for the optimal are then

$$\alpha_i = \begin{cases} \alpha_i & \text{if} \quad \left\{\nabla E(\alpha)\right\}_i \geqslant 0, \\ 0 & \text{otherwise}, \end{cases} \qquad i = 1, 2, \qquad (3-41)$$

and the control is again 'bang-bang' in nature.

In order to find numerical solutions to problems $(3-36) - (3-41)$ Algorithm 2 is used with appropriate modifications to the state and adjoint systems and to the definitions of $\alpha^k, \tilde{\alpha}^k$, and ∇E^k, which are now vector functions. The trapezoidal difference schemes and trapezium quadrature rules are applied to obtain the discretized algorithm, and the procedure remains stable and convergent.

Typical solution curves for the tidal power problem with both sluices and turbines are shown in Figure 6. Here $a_1 = 0.2$, $a_2 = 1.0$ and $K = 12.96$. It may be observed that for optimal power output the sluices are opened while the turbines are still operating, at the end of the power generation cycle. This result was not originally predicted. The average power output obtained is $\bar{P} = 0.1244 \ \rho g S F_0^2$, which compares to $\bar{P} = 0.0894 \ \rho g S F_0^2$ obtained with turbines only, assuming a corresponding value $a_0 = 0.2$ for the control constraint. We see that with the more sophisticated dual control system the energy generated can be

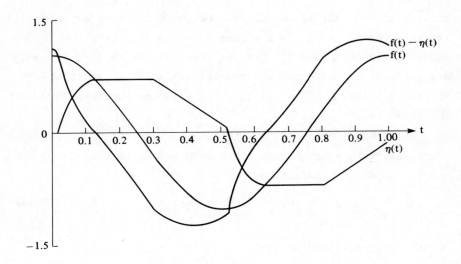

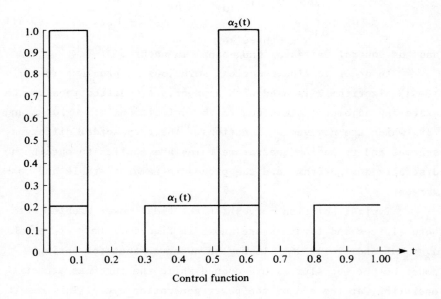

Figure 6 MODEL II — Turbines and Sluices

further increased.

An additional modification to the tidal power problem can be made in order to model ebb-generation only schemes. Reversible turbines are in general less efficient and more expensive than one-way turbines, and it is arguable that power generation only during ebb tide periods may be more cost-effective than two-way generation. In the ebb generation only case, the influx velocity is defined by

$$u_0 = \alpha_1 H(f-\eta) + \alpha_2[f-\eta] \qquad (3\text{-}42)$$

where

$$H(s) = \begin{cases} s & \text{if } s < 0 \\ 0 & \text{if } s \geqslant 0, \end{cases} \qquad (3\text{-}43)$$

and then the average power is proportional to

$$E(\alpha) = \int_0^1 \alpha_1 H(f-\eta)\, [f-\eta]\, dt . \qquad (3\text{-}44)$$

Using the conservative law as before, the optimal control problem is to maximize $E(\alpha)$ given by (3-44) subject to $\dot{\eta} = K u_0$ and $\eta(0) = \eta(1)$.

Numerical results using an appropriate modification of Algorithm 2 for the ebb generation only scheme with both turbine and sluice controls are shown in Figure 7. Here $\alpha_1 = 0.2$, $a_2 = 1.0$ and $K = 12.96$. The average energy developed by the ebb scheme is $\bar{P} = 0.0969\, \rho g S F_0^2$, which compares to $\bar{P} = 0.1244$ $\rho g S F_0^2$ from the two-way scheme. The two-way scheme thus produces about 25% more power than the ebb scheme. We observe also that for the ebb scheme the sluices are operated essentially in isolation from the turbines.

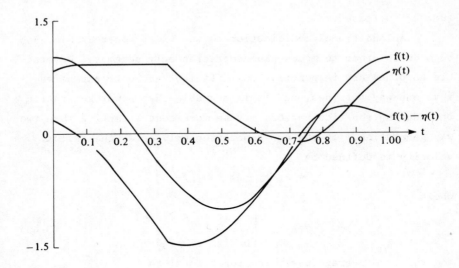

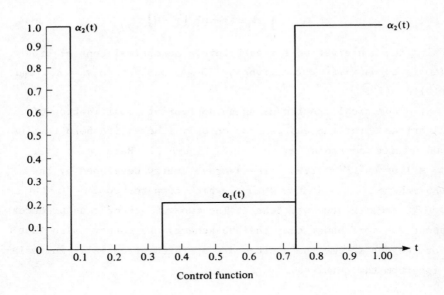

Control function

Figure 7   MODEL II — Ebb Generation

3.6 *Conclusions*

We conclude that the optimal control theory technique
developed for the linear-quadratic power generation problem associated with the oscillatory system of Section 2 can be extended
to the more general nonlinear tidal power problem and can be
applied to a number of different models representing different
energy production schemes. The results indicate that large increases in average power output can be achieved by using the
optimal time-dependent control strategies to regulate flow
through the tidal barrier.

In these models estuarine dynamics have not been included,
however, and for realistic estimates of power output from various
schemes, it is necessary to consider the effect of flow in the
tidal basin on head differences at the barrier. A simple generalization of Model II can be obtained by treating the tidal
basin as a rectangular channel which is long relative to its
depth. A one-space dimensional model of flow in the basin is
then given by the linearized shallow water equations, Stoker
[1957]. As in Model II, it is assumed that flow across the
barrier is controlled through turbines and that the influx velocity is proportional to the head difference between the surface
elevation in the basin and the tidal elevation imposed on the
seaward side of the barrier. A boundary condition, involving
the control function, is thus given at the barrier, and at the
upstream end of the basin zero flow is assumed. The flow and
surface elevation in the basin are required to be periodic in
time, and sufficient conditions to guarantee well-posedness of
the partial differential system equations are obtained (see
Kreiss [1979]). The instantaneous power generated by the flow
is again assumed proportional to the flow times the head difference at the barrier, as in Model II. The optimal control problem is then to find the control which maximizes the average power
subject to the dynamic system equations and boundary conditions.

Necessary conditions for the solution of the optimal control
problem are again derived by the Lagrangian technique. Various
new theoretical difficulties arise with the partial differential
system (see Lions [1971]), and certain modifications to the
numerical methods are required (see Ames [1977], Richtmyer &
Morton [1967]), but, in essence, the same approach used to tackle
Models I and II can be applied to the extended dynamic model.
Generalizations can also be made to include dual control of
both turbines and sluices and to simulate ebb tide generation only. Details of the techniques are described and compa-
rative results are reported in Birkett [1984] and Birkett *et al*.
[1984].

4.   CONCLUSIONS

We examine here two general models of power generation
schemes and develop techniques for determining the maximal average energy output of the schemes using optimal control theory.
The first model provides a simple test example in which power is
extracted from an oscillating system. The second model simulates
tidal power generation from flow across a tidal barrage. The
power generation problems are formulated as problems in optimal
control, necessary conditions for the optimum are given, and
numerical methods for computing solutions are developed. For
the first model, which gives a classical constrained linear-
quadratic control problem, a complete theory is derived, establishing the existence and 'bang-bang' nature of the optimal and
guaranteeing the stability and convergence of a fairly simple
numerical procedure. The same theory is derived for the second
model, but with greater difficulty. It is necessary first to
establish that the system equations are mathematically well-
posed, and, since the cost functional is no longer quadratic, a
modification must be made to the numerical iteration scheme in
order to obtain convergence. Inner iterations are also introduced in order to compute periodic solutions to the state and

adjoint equations at each step of the procedure.

Generalizations of the models to systems with multiple controls are introduced and it is demonstrated that the theoretical and numerical techniques developed can be extended to a wide class of problems. Various power generation schemes are simulated, including both two-way and ebb time generation schemes with dual controls for sluices and turbines in a tidal barrier. The principle conclusion reached is that with an appropriate control strategy the average power extracted from the generating source can be vastly increased. For the tidal power schemes, the introduction of both turbines and sluices gives an increase in power over schemes with turbines only, and two-way schemes give approximately 25% greater average energy output than ebb generation schemes (not accounting for loss of turbine efficiency). The predicted optimal strategies are not altogether obvious, especially in the dual control case, and the results of the simulations also give valuable information concerning the control policies to be adopted.

We conclude that the optimal control approach to the tidal power generation problem is a feasible and attractive method for systematically computing flow control strategies. Extensions to dynamic models show that the technique is applicable to quite complicated systems, Birkett [1984], Birkett, Count & Nichols [1984]. Further studies are now being made using refined models with more accurate data in order to obtain more realistic power output predictions. Results have already been obtained for models which incorporate dynamics in the full estuary, eliminating the assumption that the elevation on the seaward side of the barrier is unperturbed by the flow across the barrier. Nonlinear head-flow properties have also been incorporated. Details are given in Birkett & Nichols [1983]. To improve the model further, nonlinear effects in the dynamic equations and two-dimensional phenomena must be taken into account. Extensions to these more realistic cases are now being made with the joint support of CEGB and SERC , and it is expected that a global optimization technique for the complete tidal power problem can be achieved by this appr~oach.

REFERENCES

Ames, W.F.   1977, *Numerical Methods for Partial Differential Equations*. 2nd ed. Academic Press.

Berry, P.E. 1982, On the use of the describing function technique for estimating power output from a tidal barrage scheme. CEGB Marchwood Engineering Laboratories, Technical Report TPRD/M/1292/NB2 TF425.

Birkett, N.R.C.   1980, A study of an optimal power extraction problem. M Sc Thesis, University of Reading.

Birkett, N.R.C.   1984, Optimal control of dynamic systems with switches. Ph.D Thesis, University of Reading.

Birkett, N.R.C., Count, B.M., & Nichols, N.K.   1984, Optimal control problems in tidal power. *J. of Dam Construction and Water Power*. Jan. issue, 37-42.

Birkett, N.R.C. & Nichols, N.K.   1983. The general linear problem of tidal power generation with nonlinear headflow relations. University of Reading, Department of Mathematics, Numerical Analysis Report NA3/83.

Bryson, A.E. & Ho, Y.C.   1975, *Applied Optimal Control*. Halsted Press.

Count, B.M.   1980, Tidal power studies at M.E.L. CEGB Marchwood Engineering Laboratories, Technical Report MM/MECH/TF257.

Gelfand, I.M. & Fomin, S.V.   1963, *Calculus of Variations*. Prentice-Hall.

Gruver, W.A. & Sachs, E.   1980, *Algorithmic Methods in Optimal Control*. Pitman.

H.M.S.O. 1981, Tidal power from the Severn estuary. Energy Paper No. 46.

Jefferys, E.R.   1981, Dynamic models of tidal estuaries, *Proc. of BHRA 2nd Int. Conf. on Wave and Tidal Energy*.

Keller, H.B.   1968, *Numerical Methods for Two-Point Boundary Value Problems*. Blaisdell.

Kreiss, H.O.   1979, *Numerical Methods for Partial Differential Equations*. (Ed. by S.V. Parter), Academic Press.

Kreyszig, E.   1978, *Introductory Functional Analysis*. John Wiley & Sons.

Lambert, J.D.   1973, *Computational Methods in Ordinary Differential Equations*.   John Wiley & Sons.

Lee, E.B. & Markus, L.   1967, *Foundations of Optimal Control Theory*.   John Wiley & Sons.

Lions, J.L.   1971, *Optimal Control of Systems Governed by Partial Differential Equations*.   Springer-Verlag.

Pontryagin, L.S., Boltyanskii, V.G. & Gamkrelidze, R.V.   1962, *The Mathematical Theory of Optimal Processes*.   Interscience.

Richtmyer, R.D. & Morton, K.W.   1967, *Difference Methods for Initial Value Problems*.   2nd ed.   Wiley Interscience.

Stoker, J.J.   1957, *Water Waves*.   Wiley Interscience.

Wilson, E.M. *et al.*   1981,   Tidal energy computations and turbine specifications.   Institute of Civil Engineers Symposium on the Severn Barrage.

4

Low thrust satellite trajectory optimization

L. C. W. DIXON, S. E. HERSOM, AND Z. A. MAANY

1.   INTRODUCTION

In this chapter we will be concerned with finding that trajectory from earth to an asteroid, which can be obtained for a particular satellite/power unit combination using least fuel. In particular we will assume that the satellite has been launched from Earth upon an Arianne launch vehicle. The initial position at the beginning of this study will be assumed to be at a point at which the Earth's gravitational field can be neglected compared to the Sun's and will be approximately optimum within the capabilities of the launch vehicle.

The following notation will be used. The position of the satellite will be given by a vector R with origin at the Sun, the radial distance will be denoted by r, the velocity by $\dot{R}$ and the satellite mass by m.

At the starting point of the trajectory, the time t and the values of $R, \dot{R}$ and m are assumed known. It will also be assumed that the power available in an Earth orbit is P_0 and that the power available for thrust purposes at a radius r is given by $\eta P_0 r^{-k}$ where η is an efficiency factor and k is a constant ($k = 0$ for chemical systems, k = 1.7 for solar powered systems). Hence the maximum thrust available, $T_{\max}$, is given by

$$T_{\max} = 2 \eta P_0 r^{-k} / g I \; ,$$

90

where I is the specific impulse and g the gravitational constant.

The equations of motion are given by

$$\ddot{R} = -\frac{\mu R}{r^3} + \frac{T}{m}\, u \, , \qquad\qquad (1-1)$$

where T is the thrust level and u the direction of the thrust,

$$\dot{m} = -\, T/g\, I \qquad\qquad (1-2)$$

and

$$0 \leqslant T \leqslant T_{\max} = 2\eta\, P_0\, r^{-k}/g\, I\, . \qquad\qquad (1-3)$$

As the satellite is required to enter an asteroid's orbit, the final values of its position and velocity must be those associated with that orbit. Relations between $R(t_f)$ and $\dot{R}(t_f)$ are therefore known and t_f may either be prescribed or treated as an optimization variable.

The objective is to minimize the fuel used:-

$$m(t_0) - m(t_f)\, . \qquad\qquad (1-4)$$

The problem described above is the classical optimal trajectory problem, Bryson & Ho [1969], Pontryagin $et\ al$ [1962] with only one unusual feature, namely, the variation of the power available with the radius vector.

Many widely differing approaches to the solution of the classical problem have been suggested in the literature. These are divided into two distinct classes. Indirect methods attempt to find the exact solution by assuming that $T(t)$ and $u(t)$ are optimization variables and then convert this infinite dimensional problem to a finite dimensional problem by applying Pontryagin's Maximum Principle or the calculus of variations. In the early literature (c. 1970) it was noted that this approach lead to difficult optimization problems which frequently could not be solved by the available codes.

To overcome this difficulty many alternative formulations were proposed. One set of methods, termed direct methods, restricted the class of functions $u(t)$, $T(t)$ to be given functions of time containing a finite number of unknown parameters θ and then solved the resulting nonlinear programming problem in θ. A second approach was to express $R(t)$ as a given function of time which again contained a finite number of parameters θ and again solved the nonlinear programming problem in θ. Both these approaches are suboptimal in that the solution obtained depends crucially on the parameterization used; it should also be stated that if the functions and parameters are not chosen carefully then the resulting nonlinear programming problem can still be difficult; also the solution need bare little resemblance to the true optimum.

The investigation reported in this paper was undertaken to determine whether the nonlinear programming problems derived from Pontryagin's Maximum Principle would still present difficulties for the more powerful nonlinear programming codes which have been developed since 1970.

At the start of the investigation we expected that the currently available codes would easily solve the resulting nonlinear programming problem. We soon found that this expectation was not confirmed. Problems generated in this way therefore still present a challenge to researchers in nonlinear programming and in the final section of this paper we try to explain the reason for this unexpected result.

2.   MATHEMATICAL FORMULATION

As shown in the previous section, the mathematical formulation of the optimal satellite trajectory problem can be written as

$$\text{Min } m(t_0) - m(t_f)$$

subject to

$$\ddot{R} = -\frac{\mu R}{r^3} + \frac{T}{m}\,u \tag{2-1}$$

$$\dot{m} = -\,T/g\,I$$

$$0 \leqslant T \leqslant 2\eta\,P_0\,r^{-k}\,/g\,I = T_{\max} \tag{2-2}$$

where the initial values of $R(t_0)$, $\dot{R}(t_0)$ and $m(t_0)$ are given and constraints involving the values $R(t_f)$ and $\dot{R}(t_f)$ are known.

In this problem the optimization parameters are $T(t)$, $u(t)$ and t_f where u is a unit vector.

Pontryagin's Maximum Principle involves first introducing the vector $v = \dot{R}$ to make the system first order, and then introducing adjoint variables (Lagrange Multiples) associated with the differential equation constraints, p_v, p_R and p_m respectively.

The principle then introduces a Hamiltonian H given for this problem by

$$H = -p_m \, T/g\,I + p_R^T v + p_v^T \left(\frac{T}{m} u - \frac{\mu R}{r^3} \right) + \lambda (T - 2\eta\, P_0\, r^{-k} /g\,I) \; . \quad (2\text{-}3)$$

The adjoint equations which relate the values of the adjoint variables at different values of time, can be derived from the first order necessary conditions of the problem and are

$$\dot{p}_v = - \, p_R$$

$$\dot{p}_R = - \frac{3\mu}{r^5} \, (p_v^T R) - 2\eta\, P_0\, \lambda\, k\, r^{-k-2}\, R + \frac{\mu}{r^3}\, p_v \qquad (2\text{-}4)$$

$$\dot{p}_m = (p_v^T u)\, T/m^2 \; .$$

The additional Lagrange Multiplier λ for the inequality constraint is given by

$$\lambda = - \left((p_v^T u)/m - p_m /g\,I \right) \quad \text{when} \quad T = T_{\max} \qquad (2\text{-}5)$$

$$= \; 0 \qquad\qquad\qquad\qquad \text{otherwise.}$$

The Maximum Principle states that at the optimal solution the Hamiltonian H will have a maximum with respect to the control variables (i.e. $T(t)$, $u(t)$). The linearity of H with u, where u is a unit vector, implies that this maximum will occur when u is chosen parallel to p_v. The linearity of H with T, a scalar, implies that the maximum of H with respect to T will either occur when $T = T_{\max}$ or $T = 0$ depending on the sign of its coefficient

$$\frac{1}{m}\,(\boldsymbol{p}_v^{\,T}\boldsymbol{u}) - p_m/g\,I\,; \qquad\qquad (2\text{-}6)$$

this is termed the switching function. Full details of Pontryagin's Maximum Principle can be found in Pontryagin *et al* [1962] and many more recent texts including Dixon [1972].

The solution of the problem now involves two sets of differential equations, the state equations (2-1) and the adjoint equations (2-4). Both sets must be integrated numerically. Let us consider the stability of the state equations. When no thrusting is taking place then the solution is a pure gravitational Kepler trajectory. These are ellipses with focus at the Sun which are cyclic and are equally stable with respect to forward and backward integration. These arcs are termed coast arcs when considering satellite trajectories. The adjoint equations are also then cyclic. As we are considering low thrust trajectories, it is therefore equally valid to integrate either set of differential equations forward or backward in time.

The classic indirect method would then select initial values for p_m, $\boldsymbol{p}_R$ and $\boldsymbol{p}_v$; both sets of differential equations would be integrated forward in time, substituting $\boldsymbol{u} = \hat{\boldsymbol{p}}_v$ and $T = 0$ or $T_{\max}$ according to the sign of the switching function (2-6).

At a time t_f the trajectory would have reached a position R_f with velocity $\dot{R}_f$ and the maximum principle implies that this is the least fuel trajectory to that position/velocity/time combination. These final values of R_f, $\dot{R}_f$ would not however satisfy the requirements of the mission and the final values of the adjoint variable would similarly not satisfy the necessary conditions for a minimum subject to the constraints on R_f, $\dot{R}_f$ and t_f. The initial values of p_m, $\boldsymbol{p}_R$ and $\boldsymbol{p}_v$ must therefore be adjusted until these conditions are met; this problem is often solved by forming the sum of squares of the final conditions and minimizing this as a function of p_{m_0}, $\boldsymbol{p}_{R_0}$ and $\boldsymbol{p}_{v_0}$.

In previous papers, Dixon & Biggs [1972], Dixon

et al. [1981] and Dixon & Bartholemew-Biggs [1982], we had developed an adjoint-control transformation which enabled us to define p_{R_0} and p_{v_0} in terms of directional angles α_0, β_0, their time derivatives $\dot{\alpha}_0$, $\dot{\beta}_0$ and $S_0 = \frac{d}{dt}(p_v^T u)$. A detailed examination of the problem shows that the solution is independent of the magnitude $\|p_{v_0}\|$ and we choose to commence with $S_0 = (p_v^T u)_0 = \|p_{v_0}\| = 1$.

Our previous experience also pursuaded us not to minimise the sum of squares of the end conditions but instead to pose the equivalent problem

$$\min \left\{ m(t_0) - m(t_f) \right\} , \qquad (2\text{-}7)$$

subject to constraints on $R(t_f)$, $\dot{R}(t_f)$.

Given values of $\{\alpha_0, \beta_0, \dot{\alpha}_0, \dot{\beta}_0, S_0, p_{m_0}, t_f\}$ the differential equations can be integrated, and the final values of $m(t_f)$, $R(t_f)$ and $\dot{R}(t_f)$ obtained. This is therefore a nonlinear programming problem in 7 variables.

Early investigations with $k = 0$ indicated that the initial value of p_{m_0} largely determined the value of the switching function (2-6) and that if it is set incorrectly there is little to guide the optimization routine to an appropriate value.

Some initial runs were therefore undertaken on the assumption of continuous thrusting

$$T = T_{max} \qquad \text{not} \qquad 0 \leqslant T \leqslant T_{max} . \qquad (2\text{-}8)$$

Also in early tests the objective of the mission was relaxed from that of a rendezvous with the chosen asteroid (Vesta) to that of entry into any point of Vesta's orbit. This seemed a necessary first step as it was known that the different angular rates of rotation of earth and Vesta around the Sun restricted the launch dates on which rendezvous can be obtained efficiently, and we did not wish to obtain pessimistic optimal values by imposing an inappropriate combination.

The constraints for entry into any point of Vesta's orbit

can be formed in at least two different ways. In the first of
those considered the final state was defined by five equality
constraints:

(1)   The distance from the target plane should be zero.

(2)   The shortest projected distance from the end point   (2-9)
       to the target trajectory should be zero.

(3-5) The velocity at the end point should equal the velocity
       at that nearest point of the target trajectory.

This formulation, therefore, minimizes the objective
function (2-7) which is linear in t_f, as a function of the six
optimization variables $\{\alpha_0, \dot{\alpha}_0, \beta_0, \dot{\beta}_0, \dot{S}_0, t_f\}$ and subject to
the five equality constraints (2-9).

In the second constraint formulation considered, a dummy
time θ for the position of the asteroid along Vesta's orbit was
introduced as then the constraints can simply be expressed as

$$R(t_f) = R_v(\theta) \quad , \quad \dot{R}(t_f) = \dot{R}_v(\theta) . \qquad (2\text{-}10)$$

This formulation again minimizes the objective function (2-7),
but now it is a function of seven variables $\{\alpha_0, \dot{\alpha}_0, \beta_0, \dot{\beta}_0, \dot{S}_0,$
$t_f, \theta\}$ and is subject to the six equality constraints (2-10).

The original rendezvous mission may be obtained from this
second formulation in a straightforward way. It simply requires
the imposition of the link

$$\theta = t_f$$

which indicates that for continuous thrusting we have six optimization variables and six nonlinear equality constraints. This
emphasizes that Pontryagin's Maximum Principle has removed all
the degrees of freedom from this optimization problem which
reverts to the minimization of the sum of squares of the
constraints.

To complete the definition of the problem it is necessary
to specify the target orbit and the satellite details. These
are given in Appendix 1.

3. OPTIMIZATION ALGORITHMS

In the previous section it was shown that, by using Pontryagin's Maximum Principle, the optimal trajectory problem could be converted into the minimization of an objective function $F(x)$, $x \in R^n$ ($n = 6$ or 7) subject to a small number of equality constraints $e_i(x)$ $i = 1, \ldots, m$, ($m \leqslant n$).

This is therefore a problem in the standard form of nonlinear programming, namely

$$\underset{x}{\text{Min}} \quad F(x) \quad x \in R^n$$

$$\text{s.t.} \ e_i(x) = 0 \quad i = 1, \ldots, m \qquad (3-1)$$

and

$$h_j(x) \leqslant 0 \quad j = 1, \ldots, J .$$

Our problem is unusual in that no inequality constraints appear. Many codes have been written for the solution of the general nonlinear programming problem (3-1); it is a continuing research area. The codes currently available (1984) are a significant improvement in both efficiency and reliability on those available in the period 1960—1974 when many papers were written stating that the nonlinear programming problems generated by the Maximum Principle were difficult to solve. A comparison of the relative efficiency and reliability of such codes is contained in Schittkowski [1980]. One of the Numerical Optimization Centre codes OPXRQP (Bartholemew-Biggs [1979]) performed well in that study, and was therefore the first algorithm we applied to the problem.

3.1 *Results using the first model*

In all the algorithms investigated in this study an iteration defines a new approximation to the minimiser of F

$$x_{k+1} = x_k + a p_k \qquad (3-2)$$

by first choosing a direction p_k and then taking a step α in that direction.

In OPXRQP the choice of the direction p_k is based upon the solution of an equality constrained quadratic programming problem

$$\text{Min}_{p} \ \tfrac{1}{2} p^T B p + \nabla F^T p$$

$$\text{s.t} \ \nabla e^T p = -\frac{r}{2} \mu - e \qquad\qquad (3\text{-}3)$$

where

$$\left(\frac{r}{2} I + \nabla e^T B^{-1} \nabla e\right) \mu = \nabla e^T B^{-1} \nabla F - e \ .$$

In (3-3) the symbol ∇ denotes the gradient with respect to x; F, ∇F, e, ∇e are evaluated at x_k; r is a penalty parameter, and B a positive definite symmetric matrix which is meant to approximate the Hessian of the Lagrangian function of the problem. The definition (3-3) is simpler than that frequently stated but is correct for the case when (3-1) does not contain any inequality constraints.

Having obtained p from (3-3) the iteration is completed by taking a step a, which reduces the penalty function

$$P = F + \frac{1}{r} e^T e \ . \qquad\qquad (3\text{-}4)$$

A full specification of the algorithm also involves the method for updating the approximation B and the penalty parameter r.

When the algorithm was applied to the problem described in Section 2, using the constraint formulation (2-9), it was found that the path to the solution (i.e., the sequence x_k) was very dependent on the scaling given to the variables, constraints and objective function.   The problem is mathematically unaltered if we redefine

$$x_i^+ = (SX_i) \ x_i$$

$$e_i^+ = (SE_i) \ e_i$$

and

$$F^+ = (SF) \ F$$

where SX_i, SE_i and SF are positive scalings; but at any iteration the prediction p_k and the step a will be altered. The path also alters when the initial estimate $x^{(0)}$ is changed.

The algorithm OPXRQP was applied to the problem for ten different combinations of $x^{(0)}$, SX_i, SE_i and SF; it obtained the same solution on six occasions. The solution satisfied all the constraints within $\pm 2 \times 10^{-6}$ and had a free gradient of 6.25×10^{-9}.

Being disappointed with a failure rate of four out of ten we decided to compare the performance of other codes which had also performed well in Schittkowski's study.

The codes available to us for this purpose were

VMCON   Powell [1977]

VMCWD   Powell [1982]

and an implementation of Fletcher's Ideal Penalty Function method Fletcher [1975], modified at Hatfield for our system and termed OPIPF.

Of these codes VMCON is similar to OPXRQP on problems which do not contain inequality constraints; the direction p is obtained by solving

$$\begin{array}{ll} \underset{p}{\text{Min}} & \frac{1}{2} p^T B p + \nabla F^T p \\ \text{s.t} & \nabla e^T p = -e. \end{array} \qquad (3\text{-}4)$$

The step a is chosen to reduce a different penalty function, but the matrix B is again an approximation to the Hessian of the Lagrangian. This code was also shown to be very reliable and efficient in Schittkowski's study. However, when applied to the ten scalings of this problem it only converged to the solution found by OPXRQP on three occasions; on three other occasions it stopped with an objective function value which differed from the optimum by only a small number of kilograms, whilst it also failed completely on four occasions.

The theory of VMCWD is similar to that of VMCON except that it contains a safeguard aimed at enabling the iteration to continue if a point is found on nonlinear constraints away from the solution but from which a direction of significant descent on the penalty function is not obtained by (3-4). As this seemed to be the case at the near optimal stopping points mentioned above, it was hoped that this code would be both more

reliable and more efficient. In practice it only found the
solution on two occasions, near solutions on a further two, and
failed to find feasible points on six occasions.

In case the very different approach embodied in OPIPF
might have been more appropriate on this problem it was also
applied to the ten scalings but failed to obtain the solution on
every occasion and only obtained near solutions twice.

A table giving these results is shown in Appendix 2.

3.2   *Results using the Second Model*

Five tests were run with the second constraint formulation (2-10) to see if the simpler form of the constraints would
be more helpful. OPXRQP obtained the solution on three occasions; VMCON on one occasion with two very near optima and
VMCWD once with one near optima. All three codes failed on one
scaling. But the pattern on the other four scalings was unusual:

Scale 2; OPXRQP (0) 239/82; VMCON (0) 438/152; VMCWD   F

Scale 3; OPXRQP (0) 199/67; VMCON (−1) 137/32; VMCWD (0) 294/294

Scale 4; OPXRQP F   ; VMCON (−1) 150/40; VMCWD (−2) 28/28

Scale 5; OPXRQP (0) 450/131; VMCON F   ; VMCWD F .

Here the number in brackets indicates the short fall, measured
in kilograms in the fuel used; the following numbers are the
number of function calls and the number of gradient calls.

3.3   *The Introduction of a Coast Arc*

An examination of the thrusting program of the optimal
orbit of the above problem indicated that the mass in orbit
would almost certainly be improved if a coast arc were allowed.
Two additional optimization variables were included, namely, the
start time and duration of the coast phase. This was preferred
to using the switching function because it was felt that these
variables were more closely related to the operational problem.
Using the 'easy' scaling, the same solution was found by all

3 algorithms and gave a mass in orbit of 1221 kg which was an increase of nearly 30%. However, the total time of flight was increased from 303 to 396 days. (An initial thrust period of 54 days, followed by 238 days coast and a final thrust period of 104 days.)

For this problem the relative performance of the three algorithms was

OPXRQP (O) 118/59; VMCON (O) 319/106; VMCWD (O) 253/253.

3.4   *Rendezvous Mission*

In Section 2 it was shown that the rendezvous mission reduced to a problem with six equality constraints in six unknowns. The solution to this problem can therefore also be attempted using any code written to solve sets of simultaneous nonlinear equations. For interest we decided to compare the performance of the code OPNL from the Hatfield Optima Library with the three special purpose codes described above. OPNL is a simple safeguarded Newton method designed principally for robustness. It solves a set of simultaneous equations

$$e_i(x) = 0 \qquad i = 1, \ldots, n .$$

by solving

$$\nabla e^T p = -e$$

For the search direction p, and selecting the step α to reduce

$$E = e^T e .$$

The code contains safeguards to ensure that the direction p is effectively downhill on E and the step size is chosen in such a way that the iteration is guaranteed to enter a region with $\|\nabla E\| < \varepsilon_0$ in a finite number of steps.

Six combinations of scaling were chosen for this problem. All four codes succeeded on one scaling. All four codes failed on two. OPXRQP succeeded on four occasions, OPNL on three, VMCON and VMCWD on two each. The three OPNL failures converged to local minima of E which did not satisfy the constraint.

For the three scalings on which more than one code was successful the results were

	OPNL	OPXRQP	VMCON	VMCWD
No 4	96/49	83/42	96/25	F
No 5	93/47	35/25	98/26	39/39
No 6	97/44	47/27	F	39/39

Again no algorithm was consistently more efficient.

4.   CONCLUSIONS

The solution of the optimal satellite trajectory problem
via the indirect Pontryagin formulation still presents problems
for the current generation of nonlinear programming codes. The
solution of the problem is still heavily dependent on carefully
selecting appropriate scale factors for the variables and
constraints.

It is also dependent on choosing initial values of the
optimization variables which lie in the region of attraction
for the solution. In some sense the problem is similar to a
multi-extremal unconstrained problem in which other local minima
can attract the iteration away from the desired minimum if the
starting point is not within its region of attraction, Dixon
[1978]. For this problem the other points of attraction are
usually not feasible.

The other difficulty experienced by many of the codes
occurred when the iteration located a non-optimal point which
satisfied the nonlinear constraints accurately. In this situation the highly curved nature of the constraints prevented large
steps being taken along the linear arc, and this led to very
small steps away from the solution. This frequently triggered a
termination test in the algorithms, and sometimes the incorrect
message 'converged' appeared in the output. Although VMCWD
includes a technique specifically intended to overcome this
problem, on these tests it did not seem any less prone to failure
than the other codes.

The reliability of all the codes tested on this set of

problems was disappointing. Hence the main purpose of this chapter is to highlight the need for further research to generate a code robust enough to solve these problems efficiently. Two codes that it is anticipated will overcome this difficulty are under development. One is based on the Di Pillo-Grippo differentiable exact penalty function; Di Pillo-Grippo [1979], Dixon [1979], Dew [1984]. The second involves a quadratic approximation of the constraints at each iteration, Maany [1984].

Finally it should be emphasized that every optimal trajectory problem attempted has been successfully solved by the approach described in this chapter. These total at least 50 separate problems. The disappointment is that success is dependent on prior knowledge of an approximate solution and of appropriate scalings of the variables and constraints.

ACKNOWLEDGEMENT

The authors wish to acknowledge the financial support of the European Space Organisation, Darmstadt, whilst undertaking this reseacrch. They also wish to express their gratitude to M.J.D. Powell for providing the codes VMCON and VMCWD; and to Drs Roth, Flury and Hechler, for many fruitful discussions during the major research project of which the research described formed a relatively minor part. A full report is available, Dixon *et al*. [1983].

REFERENCES

Bartholemew-Biggs, M.C. 1979, An improved implementation of the recursive quadratic programming method for constrained minimization; Technical Report 105, Numerical Optimization Centre, The Hatfield Polytechnic.

Bryson,A.E. & Ho,Y.C. 1969, *Applied Optimal Control*, Gunn, London.

Dew, M.C. 1984, An exact penalty function algorithm for the optimization of Aeroengine Performance. (To appear).

Di Pillo, G. and Grippo, L. 1979, A new class of augmented Lagrangians in nonlinear programming. *SIAM J. Control Optim.*, **17**, 618-628.

Dixon, L.C.W. 1972, *Nonlinear Optimization*, E.U.P. London.

Dixon, L.C.W. & Biggs, M.C. 1971, The advantages of adjoint control transformations when determining optimal trajectories by Pontryagin's Maximum Principle. *J. of the Royal Aeronautical Society*, March, 169-174.

Dixon, L.C.W. 1978, Global optima without convexity, in *Design and Implementation of Optimization Software*. (Ed. H.J. Greenberg), Sijthoff and Noordhoff, The Netherlands.

Dixon, L.C.W. 1979, On the convergence properties of variable metric recursive quadratic programming methods. Technical Report 110, Numerical Optimization Centre, The Hatfield Polytechnic.

Dixon, L.C.W., Hersom, C.S. & Hersom, S.E. 1981, Orbit optimization. Technical Report 112, Numerical Optimization Centre, The Hatfield Polytechnic.

Dixon, L.C.W., Hersom, S.E. & Maany, Z.A. 1983, Low thrust orbit optimization for interplanetary missions. Technical Report 137, Numerical Optimization Centre, The Hatfield Polytechnic.

Dixon, L.C.W. & Bartholemew-Biggs, M.C. 1982, Adjoint-Control transformations for solving practical optimal control problems. *Optimal Control Applications and Methods*, **2** 365-381.

Fletcher, R. 1975, An ideal penalty function for constrained optimization. in *Nonlinear Programming 2*, (Eds. O.L. Mangasarian, R.R. Meyer & S.M. Robinson), Academic Press.

Maany, Z.A. 1984, An algorithm for highly curved constrained optimization. (To appear)

Pontryagin, L.S., Bottyanskii, V.G. & Gamkrelidze, R.V. 1962, *The Theory of Optimal Processes*. Interscience Press.

Powell, M.J.D. 1977, A fast algorithm for nonlinearly constrained optimization algorithms, *Numerical Analysis, Dundee 1977* (Ed. G.A. Watson). Lecture notes in Mathematics 630, Springer Verlag, Berlin.

Powell, M.J.D. 1982, VMCWD , A Fortran subroutine for constrained optimization. *WA4*, University of Cambridge.

Schittkowski, K. 1980, *Nonlinear Programming Codes*, Lecture Notes in Economics and Mathematical Sciences 183, Springer Verlag, Berlin.

Appendix 1

Numerical Data

Vesta's orbit was defined by the following data:

Perihelion	= 2.149908 Astronautical Units (A.U.)
Aphelion	= 2.573452 A.U.
Eccentricity	= 0.08967
Inclination	= 7.144 degrees
Right Ascension	= 103.489 degrees
Argument of Perigee	= 150.618 degrees.

In calculating the satellite trajectory, it has been assumed that $k = 0$ and

The Initial Mass	= 1500 kg
The Thrust	= 0.8 kg m/sec^2 = 0.04 kg Au/day^2
Specific Impulse	= 4000 sec
Rate of Flow	= 1.761 kg/day.

For the problem described in Section 3.1, one starting point was

$$x^{(0)} = (-140, \ -0.7, \ 19.4, \ 0.11, \ 0.001, \ 300);$$

The scalings were

$$SX_i = (0.1, \ 1, \ 0.01, \ 1, \ 100, \ 1)$$

and the optimal solution was

$$x^* = (-180.0491, \ 0.6187242, \ 2.69809, \ 1.763442,$$
$$-0.590798 \times 10^{-3}, \ 303.4678).$$

At this point the constraints were all satisfied within $\pm 2 \times 10^{-6}$ and the free gradient was 6.25×10^{-9}.

The other scalings used were

SE_i	1	1	10^3	10^3	10^3
SF	10^{-4} .				

Appendix 2

Table 1

Test Run No	Starting point	Scale				Integration Step Days'	OPXRQP	VMCON	VMCWD	OPIPF
		Variables	Constraints	Function	Constraints					
1	A	A	A	A	A	24	(0) 206/70	(0) 358/98	F	(-6.5) 400/97
2	A	B	A	B	A	3	(0) 1119/336	F	(-2.5) 27/27	F
3	A	C	A	A	A	24	F	(-4) 15/6	(-3.5) 10/10	F
4	B	A	A	A	A	3	F	F	F	F
5	B	B	A	A	A	24	F	F	F	F
6	A	B	A	A	A	24	(0) 906/260	(0) 115/31	(0) 43/43	F
7	C	B	A	A	A	24	F	F	F	F
8	C	D	A	A	A	24	(0) 1533/350	(0) 426/135	F	(-8) 1433/398
9	A	A	B	A	A	24	(0) 196/71	(-1) 117/38	F	F
10	A	B	A	A	B	24	(0) 927/267	(-0.5) (115/31)	(0) 100/100	F

Comparison between some of the optimization routines
Launch date 14610 Mean Julian Date
Optimal mass 965.451 kg.

Key for Table 1

1. Optimization Variables

Case	Starting Point			Solutions	Scale			
	A	B	C		A	B	C	D
$\dot{S}$	0.001	0	0.001	-5.908×10^{-3}	10.0	100.0	100.0	100.0
$\dot{\alpha}$	140.0	0	140.0	-180.05	0.1	10.0	10.0	0.1
$\ddot{\alpha}$	-0.07	0	-0.07	0.6187	1.0	1.0	1.0	1.0
$\dot{\beta}$	19.4	0	19.4	12.70	0.01	0.01	0.01	0.01
$\ddot{\beta}$	0.11	0	0.11	0.1763	1.0	1.0	1.0	1.0
T (days)	300	300	400	303.468	0.001	1.0	1.0	0.001

2. Constraints Scaling

A (1, 1, 10000, 1000, 1000)   B (10, 10, 10000, 10000, 10000)

3. Accuracy of the Objective Function

A = 10^{-6}, B = 10^{-7}, and C = 10^{-20}

4. Accuracy of the Constraints

A = 5×10^{-4}, B = 5×10^{-5}

5. Performance of the Codes

'F' indicates a failure.

The number in the brackets indicates the shortfall in the fuel used measured in kilograms. The following numbers are the number of function calls and the number of gradient calls.

5

A wire-upwinding problem

B. BENJAMIN AND D. HANDSCOMB

1.   THE REAL PROBLEM

A problem arising in the handling of wire, between its manufacture and its despatch to customers, had been studied by BICC and was brought to our attention by way of a UCINA meeting.

Following manufacture, wire is arranged in a neat pile (as illustrated in Figure 1) and subsequently wound on to reels, the wire being chopped off at standard lengths. Naturally this winding is to be carried out reasonably quickly; the problem is that snapping or snagging occurs if the winding is too fast. The question posed was whether a maximum safe winding speed could be calculated from the gauge and elastic properties of the wire and the dimensions of the installations.

Snapping, snagging, or other unwanted effects, that could interfere with production, apparently do not occur sufficiently often to make adequate physical experiments economic. This means that although many years of experience in the industry have built up much background information, this information is insufficiently precise and detailed to describe fully the situation just before failure, so that it is not known how, why, or where on the wire, the conditions leading to a breakdown first begin to appear.

Figure 1

As a preliminary step, BICC had attempted simply to compute the shape of the wire when all was going smoothly, using the simplified model described in the next section, hoping that the results would give them some insight into where things were likely to go wrong. Unfortunately even these apparently straightforward computations ran into difficulties, which is the reason for their submitting the problem to UCINA.

2.   A SIMPLIFIED MODEL

The stack is simplified to a tightly-wound helix lying around an imaginary circular cylinder with the axis vertical. The wire rises freely from the top of this stack (A in Figure 2), then passes without any intermediate constraints through a fixed smooth vertical guide B on the axis of the stack. The wire finally winds at constant speed on the reel C.

It is assumed that the wire is inextensible, that if released from the stack it would completely straighten out under its own elasticity and also that it has no preferred direction of bending.

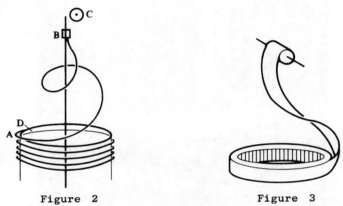

Figure 2   Figure 3

What to say about the twisting of the wire is more debatable. Presumably the wire on the stack is not twisted. Then it must acquire a twist of 2π for each complete turn of the stack that passes through the guide B. If this twist is not to build up in the segment of wire between A and B (which would certainly lead to catastrophe sooner or later), then either the wire must be wound on to the reel C in a twisted state or the axis of the reel must rotate appropriately about the axis of the helix. (The reader may like to convince himself of this by attempting to wind recording tape on to a spool from a stationary coil lying flat on the floor, as in Figure 3). With ordinary wire of circular cross-section, however, it is difficult to see how its state of twist can be controlled while it is moving.

The last remarks help to highlight the main difficulty encountered in attacking this problem, both by BICC originally

and later by ourselves: the specification of appropriate and
sufficient boundary conditions. There is no problem in describing in mathematical terms the behaviour of the wire between A
and B, as we show in the next section, but it is quite another
matter to choose conditions at A and B which not only tie in
with the physical model but also allow the resulting equations
to be solved numerically.

A final simplifying assumption that we have made is that
the wire has infinitesimal thickness. We do not believe that
anything is lost by making this assumption; it means that we
can neglect the moment of inertia of the wire about its own axis,
and also can take the vertical distance between A and B as
constant.

3.   MATHEMATICAL FORMULATION

We use the following notations:

ρ = mass of wire per unit length;

F_B = flexural rigidity of wire (resistance to bending);

F_T = torsional rigidity of wire (resistance to twisting);

g = gravitational acceleration vector

a = radius of stack;

h = height of guide B above top of stack (A);

V = speed at which wire passes through guide;

r = position vector of a point on the wire;

t = unit tangent vector;

p, b= orthogonal unit normal vectors attached to the wire
         (which would be constant if the wire were straight
         and untwisted);

K = curvature and torsion vector;

S = bending and twisting moment vector;

T = shear and tension vector;

$\cdot$ = partial differentiation with respect to time

s = arc-length along wire;

$'$ = partial differentiation with respect to s.

Between A and B the following equations are satisfied:

From the geometry:
$$r' = t ,$$
(3-1)

$$t' = K \times t , \quad p' = K \times p , \quad b' = K \times b .$$
(3-2)—(3-4)

It can be seen that, if we define $b = t \times p$, then (3-4) follows from (3-2) and (3-3); also that (3-2)—(3-4) imply that:

$t \cdot t$ is constant, as are $p \cdot p$ and $b \cdot b$;

$t \cdot p$ is constant, as are $p \cdot b$ and $b \cdot t$.

Therefore, if t, p and b are orthogonal unit vectors at any point, they will automatically be orthonormal at all points of the wire.

If we denote by ω the torsion
$$\omega = K \cdot t$$
(3-5)

then, since $t \cdot t' = 0$, it follows from (3-2) that
$$K = t \times t' + \omega t .$$
(3-6)

From elasticity
$$S = F_B (t \times t') + F_T \omega t .$$
(3-7)

From dynamics
$$S' - T \times t = 0 , \qquad T' + \rho g = \rho \ddot{r}$$
(3-8),(3-9)

Equation (3-9) assumes that r is defined relative to a fixed reference system and that s is measured from a point fixed on the wire. If the wire is moving with constant velocity V at the point B, it is useful to measure s instead from that point. In this case (3-9) becomes

$$T' + \rho g = \rho \left\{ \ddot{r} + 2V\dot{r}' + V^2 r'' \right\} .$$
(3-10)

It is also useful to transform to axes rotating with an angular velocity of V/a about the vertical axis, so that the position vector of the point A where the wire leaves the stack, as well as that of B, is independent of time when the system is in a steady state. Let Ω be a vector of magnitude V/a, directed vertically downwards. Then equation (3-10) becomes

$$T' + \rho g = \rho \left\{ \ddot{r} + 2V\dot{r}' + V^2 r'' + 2\Omega \times \dot{r} \right.$$
$$\left. + 2V\Omega \times r' + \Omega \times (\Omega \times r) \right\} ,$$
(3-11)

where r is measured from a point on the axis. In the steady state this becomes

$$T' + \rho g = \rho\left\{V^2 r'' + 2V\Omega \times r' + \Omega \times (\Omega \times r)\right\}. \qquad (3\text{-}12)$$

Equations (3-1) — (3-4),(3-7),(3-8),(3-12) are essentially those obtained by BICC, translated into vector notation. They comprise a system of six first-order ordinary differential equations and one algebraic equation in the seven vector variables r,t,p,b,K,S,T.

From these equations we may eliminate S by combining (3-7),(3-8), eliminate b, by writing $b = t \times p$, and dispense with (3-4). Such simplifications, which leave us with five ordinary differential equations in the five variables r,t,p,K and T, may however lead to awkwardness in implementing the boundary conditions.

We note that (3-2),(3-7),(3-8) imply that $\omega' = 0$, so that the torsion ω is always constant along the wire.

4.   SPECIFICATION OF EQUATIONS OF MOTION
     IN TERMS OF COMPONENTS

The equations for r', t', p', b', S' and T' can be written out in component form, yielding 18 scalar first-order nonlinear ordinary differential equations. As noted previously b, and hence b', may be removed at any time by use of $b = t \times p$, leaving 15 first-order equations.

To aid physical understanding, and thereby assist in specifying boundary conditions, we express K, S and T in terms of t, p and b. Thus we set:

$$r = x\, i^* + y\, j^* + z\, k^*, \qquad (4\text{-}1)$$
$$t = l_1 i^* + m_1 j^* + n_1 k^*, \qquad (4\text{-}2)$$
$$p = l_2 i^* + m_2 j^* + n_2 k^*, \qquad (4\text{-}3)$$
$$b = l_3 i^* + m_3 j^* + n_3 k^*, \qquad (4\text{-}4)$$
$$K = \omega t + K_2 p + K_3 b, \qquad (4\text{-}5)$$
$$S = S_1 t + S_2 p + S_3 b = F_T \omega t + F_B(K_2 p + K_3 b), \qquad (4\text{-}6)$$
$$T = T_1 t + T_2 p + T_3 b. \qquad (4\text{-}7)$$

Here k^* is a unit vector directed vertically upwards along the axis of the stack, i^* is a horizontal unit vector directed from this axis towards the point A, and j^* is the unit vector $j^* = k^* \times i^*$.

In terms of the components defined in (4-1)--(4-7), the equations (3-1) $-$ (3-4),(3-8),(3-12) for the steady state are equivalent to:

$$x' = l_1 \, , \quad y' = m_1 \, , \quad z' = n_1, \qquad\qquad\qquad \text{(4-8 a,b,c)}$$

$$l_1' = K_3 l_2 - K_2 l_3 \, , \quad m_1' = K_3 m_2 - K_2 m_3, \quad n_1' = K_3 n_2 - K_2 n_3$$
$$\text{(4-9 a,b,c)}$$

$$l_2' = \omega l_3 - K_3 l_1 \, , \quad m_2' = \omega m_3 - K_3 m_1, \quad n_2' = \omega n_3 - K_3 n_1 \, ,$$
$$\text{(4-10a,b,c)}$$

$$l_3' = K_2 l_1 - \omega l_3 \, , \quad m_3' = K_2 m_1 - \omega m_2 \, , \quad n_3' = K_2 n_1 - \omega n_3 \, ,$$
$$\text{(4-11a,b,c)}$$

$$S_1' = 0 \, , \quad S_2' = (F_B - F_T) \omega K_3 + T_3 \, , \quad S_3' = (F_T - F_B) \omega K_2 - T_2 \, ,$$
$$\text{(4-12a,b,c)}$$

$$T_1' = K_3 T_2 - K_2 T_3 + \rho\left\{ g n_1 - \Omega^2 (l_1 x + m_1 y) \right\} \qquad \text{(4-13a)}$$

$$T_2' = \omega T_3 - K_3 T_1 + \rho\left\{ g n_2 + V^2 K_3 - 2 V \Omega n_3 - \Omega^2 (l_2 x + m_2 y) \right\}$$
$$\text{(4-13b)}$$

$$T_3' = K_2 T_1 - \omega T_2 + \rho\left\{ g n_3 - V^2 K_2 + 2 V \Omega n_2 - \Omega^2 (l_3 x + m_3 y) \right\}$$
$$\text{(4-13c)}$$

where $\Omega = V/a$, and

$$S_1 = F_T \omega \, , \quad S_2 = F_B K_2 \, , \quad S_3 = F_B K_3 \, . \qquad\qquad \text{(4-14a,b,c)}$$

[Compare Love (1934), p.381.]

5.   BOUNDARY CONDITIONS

Since the length of wire between A and B is not fixed, but the relative positions in space of A and B are known, it is possibly convenient to use z (height above the top of the stack) as an independent variable in place of s. Then, for any vector $v(s)$, $dv/dz = v'/n_1$. [This change of variable is troublesome if ever $n_1 = 0$, *i.e.* if ever t is horizontal.]

Given the six first-order O.D.E.'s with z as independent

variable we could specify at $z = 0$:

$$(x,y, \quad) \quad = r(0), \qquad (l_1,m_1,n_1) = t(0) ,$$
$$(l_2,m_2,n_2) = p(0), \qquad (l_3,m_3,n_3) = b(0) ,$$
$$(S_1,S_2,S_3) = S(0), \qquad (T_1,T_2,T_3) = T(0) ,$$
$$s = s(0) = 0 ,$$

and at $z = h$:

$$(x,y, \quad) \quad = r(h) \qquad (l_1,m_1,n_1) = t(h) ,$$
$$(l_2,m_2,n_2) = p(h) \qquad (l_3,m_3,n_3) = b(h) ,$$
$$(S_1,S_2,S_3) = S(h) \qquad (T_1,T_2,T_3) = T(h) .$$

We need to select from these 35 possible conditions a set under which we can solve the equations and determine the length $s(h)$ of the wire.

However there are certain consistency requirements. As we observed earlier, the differential equations necessarily imply that $t \cdot t$, $p \cdot p$ and $b \cdot b$ are constant (say 1), as are $t \cdot p$, $p \cdot b$ and $b \cdot t$ (say 0), so that if t, p and b are specified fully at A then inconsistency or redundancy results from specifying more than three of their nine components at B. We see also from (4-12a) that S_1 is constant along the wire, so that if S_1 is given at A it must not be given at B also.

A more subtle restriction on our choice of boundary conditions arises from the following considerations. Suppose that t is specified at A by all its components, and at B by l_1, m_1, and suppose at B that $l_1^2 + m_1^2 = 1$ (which implies that $n_1 = 0$). Let J be the Jacobian matrix $\partial(l_1,m_1)/\partial(\tau_1,\tau_2)$ formed from the partial derivatives at $l_1(h)$ and $m_1(h)$ with respect to some two parameters τ_1 and τ_2, when $t(0)$ is fixed and t satisfies (3-2) for some K. Then $l_1^2 + m_1^2 + n_1^2 = 1$ for all values of τ_1 and τ_2, so that $l_1 \partial l_1/\partial \tau_j + m_1 \partial m_1/\partial \tau_j + n_1 \partial n_1/\partial \tau_j = 0$ $(j = 1,2)$. It follows that if the conditions at B are satisfied, so that $n_1 = 0$, then the Jacobian J is singular, with troublesome consequences for methods of numerical solution. Thus if we wish to specify $t(h) = (1,0,0)$, for example, we must specify

$(\ ,m_1,n_1) = (\ ,0,0)$ rather than $(l_1,m_1,\ ) = (1,0,\ )$, even though the former leaves an ambiguity of direction.

One way to help in making our selection is to think of what boundary conditions we should find physically reasonable to impose if the wire were stationary. If its length were fixed, then each end could be free, clamped, or supported in various intermediate ways. A free end corresponds to the conditions $T = 0$, $S = 0$; a clamped end to $r =$ given, $t =$ given, $p =$ given (and hence $b =$ given). An end pinned but not clamped corresponds to $r =$ given, $S = 0$ or, if held in a sleeve but free to twist $r =$ given, $t =$ given, $\omega = 0$. A more general elastic support can be defined by a system of equations of the form $T = T(r,t,p)$, $S = S(r,t,p)$. Any of these possibilities is equivalent to six scalar conditions on top of the six geometric conditions that t,p and b are somewhere orthonormal, making a total of $6 + 6 + 6 = 18$ conditions for the 18 scalar differential equations. Since physical intuition tells us that we ought to be able to determine the shape of a stationary wire from such end conditions alone (except that there may be two or more distinct possible shapes, corresponding to distinct solutions of the nonlinear equations), we should expect to be able to do so for the moving wire also.

We do, however, need one further scalar condition from somewhere in order to determine the free length of wire between A and B, presumably derived from the mechanics of what happens when the wire leaves the stack at A. We return to this point in Section 9, supposing in the meantime that the length is fixed.

6.   VARIATIONAL FORMULATION

We may get some further guidance in choosing boundary conditions by considering a variational formulation of the problem. Supposing the length of wire to be fixed, let r, t, satisfy (3-1),(3-2),(3-5), and define a quasi-potential energy:

$$E = \frac{1}{2} \int_B^A F_B t' \cdot t' + F_T \omega^2 - 2\rho r \cdot g$$

$$- \rho(Vt + \Omega \times r) \cdot (Vt + \Omega \times r) ds. \qquad (6\text{--}1)$$

We ask for this to be stationary under small perturbations of the wire, subject to suitable boundary conditions.

Perturb r to $r + \delta r$, t to $t + \delta t$, where $\delta r' = \delta t$ and $t \cdot \delta t = 0$; let this change E to $E + \delta E$ and ω to $\omega + \delta \omega$ (where $\omega' = \delta \omega' = 0$). Then, to first order,

$$\delta E = \int_A^B F_B \delta t' \cdot t' + F_T \omega \delta \omega - \rho \delta r \cdot g$$

$$- \rho(Vt + \Omega \times r) \cdot (V\delta t + \Omega \times \delta r) ds. \qquad (6\text{-}2)$$

If (3-7),(3-8),(3-12) are satisfied, and certain boundary conditions hold, then we may, by means of much substitution and integration by parts, obtain

$$\delta E = \int_A^B -F_B \delta t \cdot t'' + F_T \omega \delta \omega - T \cdot \delta t \, ds$$

$$= F_T \omega \int_A^B \delta \omega + (t' \times t) \cdot \delta t \, ds = 0, \qquad (6\text{-}3)$$

provided either that the two ends of the wire are prevented from twisting, so that the last integral vanishes, or that $\omega = 0$, which is the case when either end may twist freely.

The boundary conditions used in obtaining (6-3) are, at each end A and B of the wire

$$\left\{ T - \rho V(Vt + \Omega \times r) \right\} \cdot r = 0, \quad t' \cdot \delta t = 0. \qquad (6\text{-}4)$$

These are certainly satisfied if r and t are both fixed ($\delta r = \delta t = 0$), so that the end of the wire is fixed in position and direction, but we can see that there are alternative conditions under which (6-4) holds; for instance, δt need not vanish if t' vanishes at that end (whether t' is constrained to vanish or does so accidentally). We have not explored the implications of these alternatives.

7.   NUMERICAL SOLUTION OF THE TWO-POINT
       BOUNDARY-VALUE PROBLEM

The 18 scalar first-order O.D.E.'s (4-8a)—(4-13c) are to be
solved subject to boundary conditions at A and B as described in
Section 5. For numerical solution it is convenient to replace
equations for the components of S by equations for the components of K, using (4-14a-c), so that (4-12a-c) are replaced by:

$$\omega' = 0, \quad K_2' = (F_B - F_T)\,\omega K_3/F_B + T_3/F_B \, ,$$
$$K_3' = (F_B - F_T)\,\omega K_2/F_B - T_2/F_B . \qquad (7\text{-}1a,b,c)$$

The whole system of equations takes the form

$$d\boldsymbol{v}/ds = \boldsymbol{f}(\boldsymbol{v}) \, , \qquad (7\text{-}2)$$

where $\boldsymbol{f}$ is a nonlinear vector-valued function.

We have already pointed out in several places that three
of these equations, say those (4-11 a,b,c) corresponding to
$\boldsymbol{b}' = K \times \boldsymbol{b}$, can be removed from the system by using $\boldsymbol{b} = \boldsymbol{t} \times \boldsymbol{p}$; *i.e.*

$$l_3 = m_1 n_2 - m_2 n_1 \, , \text{ etc.} \qquad (7\text{-}3)$$

This reduction in the number of equations and dependent variables
has not been exploited to date because the resulting equations
lose some symmetry and, while this alone is not too serious, it
has the effect that the Jacobian matrix $\{J_{ij} = \partial f_i/\partial v_j\}$, used in
the numerical solution, is less sparse and the elements are more
complicated.

These arguments apply even more strongly to further reduc-
tions that might be possible, such as eliminating the equation
for n_2 by using

$$n_2^2 = 1 - l_2^2 - m_2^2 \qquad (7\text{-}4)$$

and that for n_1 by using

$$n_1 = -(l_1 l_2 + m_1 m_2)/n_2 . \qquad (7\text{-}5)$$

We therefore attempt to solve (4-8a) — (4-11c), (4-13a-c),
(7-1a-c) in the form stated.

It is instructive to discuss first a special case for
which there are known solutions. If we set

$$\boldsymbol{r} = (a\cos\lambda s, \; a\sin\lambda s, \; c\lambda s) \, , \qquad (7\text{-}6)$$

so that r is the position of a point on a right circular helix $(a^2 + c^2 = \lambda^{-2})$, then the functions

$$
\left.
\begin{aligned}
t &= (-\lambda a \sin \lambda s, \ \lambda a \cos \lambda s, \ c\lambda), \\
p &= (-\cos \lambda s, \ -\sin \lambda s, 0), \\
b &= (c\lambda \sin \lambda s, \ -c\lambda \cos \lambda s, \ a\lambda), \\
T &= (0,0,0), \\
K &= (0,0,\lambda), \ \left\{ \text{in components of } t,p,b \right\} \\
S &= (0,0,F_B \lambda),
\end{aligned}
\right\}
\tag{7-7}
$$

satisfy the system

$$
r' = t, \quad t' = K \times t, \quad p' = K \times p, \quad b' = K \times b,
$$
$$
S' = T \times t, \quad T' = 0,
\tag{7-8}
$$

which, we note, are a special case of our original steady-state equations with $V = 0$ and $g = 0$. When we tried to solve (7-8) subject to

$$
r(0) = (0.5, 0, 0), \quad t(0) = (0, 0.96, 0.28),
$$
$$
p(0) = (-1, 0, 0),
\tag{7-9} - (7-11)
$$
$$
b(0) = (0, -0.28, 0.96), \ S(0) = (0, 0, F_B \lambda),
$$
$$
T(0) = (0,0,0),
\tag{7-12)-(7-14}
$$

we had no trouble with standard initial value library programs, producing answers that checked well against the known solution (7-6), (7-7). Replacing condition (7-14) at $s = 0$ by

$$
r(25\pi/24) = (0.5, 0, 7\pi/24)
\tag{7-15}
$$

we get a two-point boundary-value problem. Application of a multiple shooting method from I.M.S.L., or a finite-difference method with deferred correction (essentially Pereyra's PASVA 3; see Pereyra [1979]) from the N.A.G. Library, succeeded in each case in producing the desired results.

If we set

$$
r = (R \cos \theta, \ R \sin \theta, \ c\theta),
$$

where

$$
R = a \cos n\theta,
$$

then it is possible to generate solutions which seem closer in shape to the wire in Section 2, representing a curve passing through points A and B. The solution is particularly simple when $n = 1$, when we have

$$r = (a \cos \lambda s \cos \lambda s,\ a \cos \lambda s \sin \lambda s,\ c \lambda s),$$
$$t = (-a \lambda \sin 2 \lambda s,\ a \lambda \cos 2 \lambda s,\ c \lambda),$$
$$p = (-\cos 2 \lambda s,\ -\sin 2 \lambda s,\ 0),$$
$$b = (c \lambda \sin 2 \lambda s,\ -c \lambda \cos 2 \lambda s,\ a \lambda),$$
$$K = (0, 0, 2 \lambda),$$
$$T = (0, 0, 0),$$
$$S = (0, 0, 2 F_B \lambda).$$

$$(7\text{-}16)$$

Functions (7-16) also satisfy the differential equations (7-6) — (7-7), and when we tried to solve these equations subject to

$$r(0) = (0.5, 0, 0),\ t = (0, 0.3655, 0.9308),\ p(0) = (-1, 0, 0),$$
$$b(0) = (0, -0.9308, 0.3655),\ T(0) = (0, 0, 0),\ S(0) = (0, 0,\ )$$

and

$$r(2.1487) = (0, 0, 2),\ t(2.1487) = (\ , -0.3655,\ ),$$

then the two-point boundary-value routines mentioned above again seemed to produce the correct results without difficulty.

However, when we now tried to solve similar problems with $g \neq 0$, difficulties arose. For instance, a continuation scheme was tried, starting with the known solution for $g = 0$, then gradually increasing the value of g using the solution at one step as the initial estimate required for the next. As g increased, the codes showed increasing signs of distress and eventually threw in the towel. (We have had limited success with this scheme when the length of wire was much shorter, however, which could possibly indicate the existence of a critical length.)

This represents the stage at which our numerical investigations have remained for some time despite all our efforts. The difficulty is principally that we can see no way of deciding (in a nonlinear problem of this complexity) whether failure to reach a solution is due to a defect (not necessary a positive blunder) in the detailed coding of the library procedures, to numerical instability when such algorithms are applied to the present

equations, or to inherent instability in the equations themselves — and in the last event whether the instability has a
physical meaning or is merely due to the way the equations are
formulated.

8.   NUMERICAL SOLUTION BY VARIATIONAL METHODS

We should report that we have recently had some promising
results from applying a constrained minimization technique to the
variational formulation of Section 6, after a little transformation.

Briefly the scheme is the following. We represent the
wire by a chain of finite links of length ds, the orientation
of the ith link being given by the unit vectors t_i, p_i and b_i
and its ends by r_{i-1} and r_i. We write the three unit vectors
in terms of four scalar parameters as follows:

$$t_i = \left(2(b_i d_i + a_i c_i) , \quad 2(c_i d_i - a_i b_i), \quad a_i^2 - b_i^2 - c_i^2 + d_i^2 \right),$$
$$p_i = \left(a_i^2 + b_i^2 - c_i^2 - d_i^2 , \quad 2(b_i c_i + a_i d_i), \quad 2(b_i d_i - a_i c_i) \right),$$
$$b_i = \left(2(b_i c_i - a_i d_i) , \quad a_i^2 - b_i^2 + c_i^2 - d_i^2, \quad 2(c_i d_i + a_i b_i) \right).$$

$$(8-1)$$

These represent orthonormal vectors if and only if

$$a_i^2 + b_i^2 + c_i^2 + d_i^2 = 1 . \qquad (8-2)$$

We can then write down a discrete approximation to E, (6-1), in
terms of r_i, a_i, b_i, c_i and d_i; we can also use these variables
to represent the boundary conditions. We then search for a minimum of E subject to the constraints of the boundary conditions,
of (8-2), and of

$$r_i - r_{i-1} = t_i \, ds . \qquad (8-3)$$

If there are n links in the chain, this means minimizing a function of $O(7n)$ variables subject to $O(4n)$ constraints.

We have successfully solved a static problem $(V = 0$ but
$g \neq 0)$ by these means, taking $n = 10$, and hope that the method
will extend to the dynamic problem without difficulty, at least

until the winding speed approaches the danger level.

9.   THE FREE BOUNDARY

There are many ways in which the position of the point A might be defined, different physical assumptions leading to different boundary conditions. One such is the following.

Assume that the wire comes off the stack 'cleanly' at A; that is, assume that all constraints are removed instantaneously and simultaneously there. [This is not necessarily the case in reality; the wire may slide across the top of the stack before it finally leaves it, or it may ride up the sides of the supporting cage (see Figure 1) for a while before moving in towards the axis.] Let D (Figure 2) be a point on the stacked portion of the wire, such that the length of wire between D and B is fixed, and redefine E in (6-1) so that the integral runs from D to B. Then we ask for E to be stationary under small perturbations in the wire between A and B and also under small perturbations of the distance DA, subject to the conditions that the wire is in its original coiled form between D and A, that r, t, p and b are continuous at A, and that the conditions at B are as before.

The additional condition that E be stationary with respect to small changes in DA will be satisfied if the integrand is continuous at A. Since we already have r and t continuous there, it is enough to say that

$$F_B t' \cdot t' + F_T \omega^2 \qquad (9\text{-}1)$$

should be continuous at A. With the assumptions we have made in Section 2 about the state of the wire on the stack, this gives the scalar boundary condition to be imposed at A:

$$F_B t' \cdot t' + F_T \omega^2 = F_B / a^2 \qquad (9\text{-}2)$$

or

$$F_B (K_2^2 + K_3^2 - 1/a^2) + F_T \omega^2 = 0. \qquad (9\text{-}3)$$

The difficulties we have encountered in solving the differential equations with fixed boundaries have prevented us from exploring this aspect of the problem any further. We may, how-ever, be able to build this feature into the variational approach, simply by introducing additional constraints.

We have had helpful discussions with many colleagues in the course of this investigation, among whom we particularly wish to mention Professors L. Fox and S.S. Antmann, Drs J.S. Rollett and J.R. Ockendon, and Mr P. Heyda of BICC. We must above all acknowledge the encouragement we have had from Dr Sean McKee at all stages.

REFERENCES

Antmann, S.S. 1980, Nonlinear eigenvalue problems for the whirling of heavy elastic strings. *Proc. Roy. Soc. Edinb.* **85A**, 59–85.

Antmann, S.S. & Kenny, C.S. 1981, Large buckled states of nonlinearly elastic rods under torsion, thrust and gravity. *Arch. Rat. Mech. and Anal.* **76**, 289–338.

Love, A.E.H. 1934, *A Treatise on the Mathematical Theory of Elasticity* (4th ed.) Cambridge University Press.

Pereyra, V. 1979, An adaptive finite-difference Fortran program for first-order nonlinear ordinary boundary problems. In *Springer Lecture Notes in Computing*, No. 76.

Stuart, C.A. 1975, Spectral theory of rotating chains. *Proc. Roy. Soc. Edinb.* **73A**, 199–214.

6

Analysis of a wave power device

B. M. COUNT AND C. M. ELLIOTT

1.   INTRODUCTION

The theoretical analysis of structures moving under the influence of waves has to date been mostly linear and has proved adequate for a large number of calculations on ships. Wave power devices have been subjected to similar analysis where the power extraction mechanism is linear and results have correlated extremely well with experimental data indicating that linear hydrodynamic theory is adequate under normal operating conditions, Count [1977]. The theory has been invaluable in understanding the physics of wave energy absorption, but for more detailed engineering design data the consequences of employing practical nonlinear power take-offs will need to be considered.

Engineering studies of the methods for extracting energy from waves have identified the technical feasibility of using both hydraulic machines and air turbines for power conversion. This equipment will not, of course, be able to provide a perfect linear load, although in certain situations a reasonable approximation to linearity may be obtained. However it is anticipated that more practical systems will have a substantial degree of nonlinearity.

In addition there are economic and practical incentives to pneumatically or hydraulically couple many wave energy devices

to share either a single turbine or, possibly, two turbines. Not
only does this serve to reduce the maintenance costs of this part
of the system by reducing the number of machines, but also the
variability of the delivered power is reduced because each device
will deliver its power at different phases. It is important,
however, to consider the effect on the hydrodynamic behaviour
brought about by such an arrangement since a dramatic reduction
in primary efficiency may outweigh the practical advantages.

In order to model a manifold system of the type described
above, it may be assumed that, to a first approximation, the flow
(of air or fluid) through the turbine is constant. It follows
that each device will be subjected to a constant resistive force
analogous to Coulomb friction. This type of power extraction is
nonlinear because its mathematical form is

$$\varepsilon \, \text{sign} \, (\dot{x})$$

where $\varepsilon (>0)$ is the resistive force and $\dot{x}$ is the speed of the
device.

It will be assumed that linear hydrodynamic theory can be
used. This may be justified on the basis that any sensible wave
power system needs to extract energy efficiently in modest
(linear) wave conditions and will be less efficient in extreme
(nonlinear) wave conditions. Indeed, in violent conditions there
is little doubt that the power transferred from the wave to the
device will be far in excess of the requirement and that survival
will be the major concern. At modest power levels the hydrodynamic characteristics are well approximated by linear theory.
Within this range performance will be important and the dominant
nonlinear effects will be those of the power extraction system.

In this chapter the general equation of motion of any
structure excited by waves is derived assuming linear hydrodynamic theory. Under these assumptions a nonlinear integro-
differential equation is derived representing a motion, forced
by incident waves, and with the motion induced forces accounted
for by a convolution integral. The kernel of this integrand is

directly related to the added mass and damping coefficients which have previously been derived, Count [1982]. The mathematical problem is to solve the initial value problem for the equation

$$m\ddot{x} + \varepsilon \, \text{sign}\,(\dot{x}) + Bx + \int_0^t G(t-\tau)\,\dot{x}(\tau)\,d\tau = F(t)\,. \qquad (1\text{-}1)$$

The discontinuous nature of the nonlinearity gave rise to numerical difficulties and the first author brought this aspect of the problem to the attention of the Oxford Study Group with Industry in 1978. The second author proposed a novel numerical method and analysed it for a wide class of ordinary differential equations, Elliott [1985]. The method proved to be highly success-ful in the context of this study. Subsequently the analysis was extended to integro-differential equations in Elliott & McKee [1981].

The layout of the chapter is as follows. In Section 2 the integro-differential is derived. A numerical method is des-cribed in Section 3. The application of the numerical method is discussed in Section 4. Some conclusions are reached in Section 5.

2.   FORMULATION OF THE WAVE ENERGY PROBLEM

2.1   *The Equation of Motion*

For simplicity a wave power device constrained to move with a single degree of freedom is analysed when connected to a nonlinear extraction mechanism. The device will be assumed to have inertia, I, hydrostatic stiffness, B, and is excited by in-cident waves which are described by linear hydrodynamics, Count [1982]. The equation of motion can be written as:

$$I\ddot{x} + P(x,\dot{x},\ddot{x}) + Bx = F_{w(t)} \qquad (2\text{-}1)$$

where $F_{w(t)}$ is the wave induced force, $P(\ddot{x},\dot{x},\ddot{x})$ characterises the energy extraction mechanism and x denotes the body displace-ment.

This equation has been solved for linear forms of $P(x,\dot{x},\ddot{x})$, Count [1977], where the wave forces were calculated from the full linear hydrodynamic equations. Using the partitioning described

by Newman [1962] the wave forces were separated into two components; an exciting force $F(t)$, say, which is due to an incident
wave upon a fixed structure and a radiation force induced by the
body motion. In the case of linear loading this representation
is particularly convenient since equation (2-1) simply describes
a forced oscillation of a mass-spring-damping system which is
easily solved. This partitioning is still useful for nonlinear
device motions, although the resulting equation is more complex
than in the linear case and the fluid equations must be examined
in more detail.

2.2   *The Fluid Equations*

The linear equations for incompressible, irrotational,
inviscid flow are in terms of a velocity potential $\Phi(\boldsymbol{r}, t)$ satisfying Laplace's equation

$$\nabla^2 \Phi = 0$$

where $\boldsymbol{r} = (X_1, X_2, X_3)$ and from which the fluid velocities $\boldsymbol{V}(\boldsymbol{r}, t)$
are calculated as

$$\boldsymbol{V}(\boldsymbol{r}, t) = -\nabla \Phi.$$

The standard linearized free surface boundary condition equating
the velocity of the fluid to that of the surface displacement is,
with X_2 positive downwards into the fluid,

$$\frac{\partial \Phi}{\partial X_2} - \frac{1}{g} \frac{\partial^2 \Phi}{\partial t^2} = 0 \quad \text{on } X_2 = 0, \tag{2-2}$$

and the continuity condition applied on the mean wetted surface
of the body is

$$-\frac{\partial \Phi}{\partial n} = V_n(\boldsymbol{r}, t), \tag{2-3}$$

where n denotes the outward normal direction and $V_n(\boldsymbol{r}, t)$ is the
normal velocity of the immersed body boundary. Further necessary
conditions are that at large distances from the body, simple waves
shall exist and the potential decays to zero as X_2 increases
(Stoker [1957]). Having solved for Φ, the dynamic pressure is
given by Bernoulli's Theorem (Stoker [1957]) as

$$P = -\rho \frac{\partial \Phi}{\partial t},$$

and by evaluating this around the body the net forces acting on it can be calculated.

2.3   *A Convenient Partition*

As a consequence of the linearization the solution can be separated into any number of superposable parts satisfying Laplace's equation. It is convenient to distinguish three such potentials, say Φ_0, Φ_d and Φ_R. The potential Φ_0 will be that due to an incident plane wave alone in the absence of a body, Φ_d will be the perturbation on the incident wave due to a body at rest and is known as the diffraction potential, and Φ_R is the potential due to forced movement of the body in nominally calm water and is known as the radiation potential. The overall solution is then, Newman [1962],

$$\Phi = \Phi_0 + \Phi_d + \Phi_R \, .$$

This particular choice of separate potentials is convenient since the wave induced force acting on the body is simply the sum of an exciting force, due to Φ_0 and Φ_d, and a force proportional to the body motion described by Φ_R. If n_x denotes the component of the normal vector in the direction of motion, x, then the exciting force, $F(t)$, is

$$F(t) = -\rho \int_{body} \frac{\partial(\Phi_0 + \Phi_d)}{\partial t} \, n_x \, dS \, ,$$

which can be determined by solving the hydrodynamic equations in the frequency domain and transforming back to the time domain. The induced forces are

$$-\rho \int_{body} \frac{\partial \Phi_R}{\partial t} \, n_x \, dS \, ,$$

where the potential Φ_R is only known for given motion amplitudes at any particular frequency. Mathematically the Fourier transform of Φ_R can be determined which must be converted back into the time domain.

2.4   *The Motion Induced Force*

If the equations of Section 2.2 are transformed to the frequency domain, only the boundary conditions (2-2) and (2-3) are altered to

$$\frac{\partial \tilde{\Phi}_R}{\partial X_2} + K\tilde{\Phi}_R = 0 \quad \text{on} \quad X_2 = 0,$$

and

$$-\frac{\partial \tilde{\Phi}_R}{\partial n} = \tilde{V}_n(\boldsymbol{r}, \omega),$$

where $K = \omega^2/g$, $\tilde{\Phi}_R$ is the Fourier transform of Φ_R and $\tilde{V}_n$ the transform of V_n. The transformed fluid equations have been solved, Count [1982], and in general the force

$$i\omega\rho \int_{\text{body}} \tilde{\Phi}_R n_x dS$$

is expressed as

$$\left[-N(\omega) + i\omega M(\omega) \right] \tilde{V}(\omega), \qquad (2\text{-}4)$$

where $N(\omega)$ and $M(\omega)$ are the added damping and mass parameters respectively. The term $\omega M(\omega)$, however, is divergent as $\omega \to \infty$ since the added mass, $M(\omega)$, tends to a finite value. This physically represents the instantaneous effect of the fluid on the body. Expression (2-4) is therefore re-written as

$$\left[-N(\omega) + i\omega(M(\omega) - M_\infty) \right] \tilde{V}(\omega) + i\omega M_\infty \tilde{V}(\omega),$$

where M_∞ is the high frequency value of $M(\omega)$, since the velocity V is $\dot{x}(t)$ and the expression for the motion induced force is a product of Fourier transforms. In the time domain this can be expressed as, Cummings [1962],

$$-\int_0^t G(t-\tau) \, \dot{x}(\tau) \, d\tau - M_\infty \ddot{x}(t), \qquad (2\text{-}5)$$

where the convolution for the second term is considerably simplified. The integration limit of the first term is a direct consequence of causality conditions since no force can depend on future behaviour of the structure. The function $G(t)$, is given by the relation

$$\int_0^\infty G(t) \, e^{i\omega t} \, dt = N(\omega) - i\omega(M(\omega) - M_\infty),$$

giving the real transformation,

$$G(t) = \frac{2}{\pi} \int_0^{\infty} N(\omega) \cos \omega t \, d\omega . \qquad (2\text{-}6)$$

The above two equations are consistent since

$$N(\omega) \quad \text{and} \quad -\omega(M(\omega) - M_{\infty})$$

are related via the Kramers-Kronig relations, Kotik and Mangulis [1962].

The general equation of motion can therefore be written as the integro-differential equation

$$(I + M_{\infty})\ddot{x} + P(x, \dot{x}, \ddot{x}) + Bx + \int_0^{t} G(t - \tau) \, \dot{x}(\tau) \, d\tau = F(t) \quad (2\text{-}7)$$

where $G(t)$ is determined from the added damping parameter and $F(t)$ is determined from the nature of the incident waves. It is interesting to note that as a result of the Kramers-Kronig relations, knowledge of the added damping $N(\omega)$ and M_{∞} is sufficient to determine the motion induced forces.

The particular case of interest is when

$$P(x, \dot{x}, \ddot{x}) = \varepsilon \operatorname{sign}(\dot{x}), \qquad \varepsilon > 0 \qquad (2\text{-}8)$$

which simulates either a single hydraulic pump discharging into a constant reservoir or the resistance seen by a single device pneumatically coupled to many others. It turns out that not only does $\ddot{x}$ jump discontinuously as $\dot{x}$ instantaneously changes sign, but also there exists the possibility that motion ceases for an interval of time. In order to see this we interpret (2-7) — (2-8) as the problem of motion of a block on a rough plane subject to a continuous applied force. The equation of motion is

$$(I + M_{\infty})\ddot{x} + \mathfrak{F} = f(t) \qquad (2\text{-}9)$$

where $f(t)$, (suppressing the dependence on x and $\dot{x}$), is the applied force and $\mathfrak{F}$ is a frictional force modelled by Coulomb (or dry) friction, Den Hartog [1930, 1931]. Thus during motion, $\dot{x} \neq 0$, $\mathfrak{F}$ takes its limiting value, ε, and opposes the motion so that

$$\dot{x} \neq 0 \Rightarrow \mathfrak{F} = \varepsilon \operatorname{sign}(\dot{x}) . \qquad (2\text{-}10)$$

Let $z(t^*+0)$ and $z(t^*-0)$ denote values of $z(t)$ in the intervals $(t^*, t^*+\delta)$ and $(t^*-\delta, t^*)$ for some $\delta > 0$. If $\dot{x}(t^*) = 0$ and $\dot{x}(t^*-0) \neq 0$ then *either* the motion may continue so that $\dot{x}(t^*+0) \neq 0$ and

$$(I+M_\infty)\ddot{x}(t^*+0) = f(t^*+0) - \varepsilon\ \text{sign}\ x(t^*+0) \neq 0 \qquad (2\text{-}11)$$

and

$$|f(t^*+0)| > \varepsilon$$

or

$$|f(t^*+0)| \leqslant \varepsilon \quad \text{and} \quad \dot{x}(t^*+0) = \ddot{x}(t^*+0) = 0, \quad \mathfrak{F} = f(t^*+0). \ (2\text{-}12)$$

In the case (2-11) $\ddot{x}$ changes discontinuously at t^*, provided $\dot{x}$ changes sign, and in the case (2-12) there is an interval of time in which the applied force is unable to overcome the frictional force and motion ceases. Indeed the block will remain stationary until the applied force is greater than the limiting frictional force. This type of behaviour of $\dot{x}$ is called stiction. It is well known in the theory of mechanical oscillations see Den Hartog [1930, 1931], Andronov, Vitt & Khaikin [1966] and Threfall [1978].

When standard ordinary differential equation solvers are used to solve these equations without special provision being made for the stiction intervals, they frequently jump to and fro across the surface $\dot{x} = 0$. This 'chattering' can cause the program to be very slow or grind to a halt. Indeed it was an experience of this kind which lead the first author to bring this numerical problem to the Oxford Study Group with Industry.

3.   THE NUMERICAL METHOD

Let $\text{sign}(r)$ be considered as a multivalued mapping at $r = 0$ and be defined by

$$\text{sign}(r) \begin{cases} = 1 & r > 0 \\ \in [-1, 1] & r = 0 \\ = -1 & r < 0 . \end{cases} \qquad (3\text{-}1)$$

Since the frictional force $\mathfrak{F}$ always satisfies

$$\mathfrak{F} \in \varepsilon\ \text{sign}(\dot{x}), \qquad (3\text{-}2)$$

equation (2-7) implies that

$$(I + M_\infty)\, \ddot{x} + Bx + \int_0^t G(t - \tau)\, \dot{x}(\tau)\, d\tau - F(t) \in -\varepsilon\, \text{sign}(\ddot{x})\,. \quad (3\text{-}3)$$

Setting $y = \dot{x}$, (3-3) becomes the first order system

$$\dot{x} - y = 0 \qquad\qquad\qquad (3\text{-}4)$$

$$(I + M_\infty)\, \dot{y} + Bx + \int_0^t G(t - \tau)\, y(\tau)\, d\tau - F(t) \in -\varepsilon\, \text{sign}(y)\,.$$

We call (3-4) an ordinary integro-differential inclusion.
Since $\dot{y}$ may have jump discontinuities we can only demand that
(3-4) holds for almost all t in the interval $(0, T)$. Thus the
initial value problem is to find $x(t)$ and $y(t)$ such that:

(i) $x(t)$ and $y(t)$ are Lipschitz continuous on $[0, T]$

(ii) $x(0)$ and $y(0)$ are prescribed

(iii) (3-4) holds for almost all t in $[0, T]$.

It is known that the initial value problems for differential in-
clusions of the form (3-4) have unique solutions. Furthermore
the right-hand derivative

$$\frac{d^+ y(t)}{dt} = \lim_{\substack{h \to 0 \\ h > 0}} \frac{y(t + h) - y(t)}{h}$$

exists for all t, and is given by the formula

$$\frac{d^+ y(t)}{dt} = \left(F(t) - Bx - \int_0^t G(x - \tau)\, y(\tau)\, d\tau - \varepsilon\, \text{sign}(y) \right)^0\,. \quad (3\text{-}5)$$

For any r and b we use the notation that $(b - \varepsilon\, \text{sign}(r))^0$ is
that element of $(b - \varepsilon\, \text{sign}(r))$ with least magnitude. Therefore,
for $y \neq 0$,

$$\frac{d^+ y}{dt} = F(t) - Bx - \int_0^t G(t - \tau)\, y(\tau)\, d\tau - \varepsilon\, \text{sign}(y)$$

and for $y = 0$,

$$\frac{d^+ y}{dt} = 0$$

when

$$\left| F(t) - Bx - \int_0^t G(t - \tau)\, y(\tau)\, d\tau \right| \leqslant \varepsilon$$

and

$$y(t + 0) = 0\,,$$

or

$$\frac{d^{+}y}{dt} = F(t) - Bx - \int_{0}^{t} G(t-\tau)\, y(\tau)\, d\tau + \varepsilon < 0 \quad \text{and} \quad y(t+0) < 0,$$

or

$$\frac{d^{+}y}{dt} = F(t) - Bx - \int_{0}^{t} G(t-\tau)\, y(\tau)\, d\tau - \varepsilon > 0 \quad \text{and} \quad y(t+0) > 0.$$

Thus the values of $d^{+}y/dt$ given by (3-5) agree with the definition of $\ddot{x}$ in (2-9) $-$ (2-12) and the inclusion (3-4) is equivalent to our original problem. The numerical method for approximating $x(t)$ is obtained by discretising (3-4). Let x_n and y_n be approximations to $x(nh)$ and $y(nh)$ defined by

(a)
$$\frac{x_n - x_{n-1}}{h} - y_n = 0$$

(b)
$$(I + M_\infty) \frac{y_n - y_{n-1}}{h} + Bx_n + h \sum_{j=0}^{n} \gamma_{n,j}\, G((n-j)h)\, y_j \qquad (3\text{-}6)$$

$$- F(nh) \in -\varepsilon\, \text{sign}(y_n)$$

where $h = T/N$, and the quadrature weights are

$$\gamma_{n,0} = \gamma_{n,n} = \tfrac{1}{2}; \quad \gamma_{n,j} = 1, \quad 1 \leqslant j \leqslant n-1.$$

Substituting (3-6a) into (3-6b) leads to the inclusion

$$Q\, y_n - R_{n-1} \in -h\varepsilon\, \text{sign}(y_n) \qquad (3\text{-}7)$$

where

(a)
$$Q = I + M_\infty + h^2 (B + \tfrac{1}{2} G(0))$$

(b)
$$R_{n-1} = hF(nh) + (I + M_\infty)y_{n-1} - hBx_{n-1} \qquad (3\text{-}8)$$

$$- h^2 \sum_{j=0}^{n-1} \gamma_{n,j}\, G((n-j)h)\, y_j.$$

For sufficiently small h, at least, Q is positive hence the inclusion (3-7) has the unique solution

$$y_n = \begin{cases} (R_{n-1} - h\varepsilon)/Q > 0, & R_{n-1} > h\varepsilon \\ 0, & |R_{n-1}| \leqslant h\varepsilon \\ (R_{n-1} + h\varepsilon)/Q < 0, & R_{n-1} < h\varepsilon. \end{cases} \qquad (3\text{-}9)$$

x_n can now be evaluated by substitution into (3.6a).
When dy/dt possess a finite number of jump discontinuities in
the interval $[0, T]$, the numerical method is first order accurate, Elliott & McKee [1981] and Elliott [1985], so that,

$$|y_n - y(nh)| \leqslant ch , \qquad |x_n - x(nh)| \leqslant ch$$

where c is a constant independent of n and h.

The discrete values also have the attractive feature that
there exist discrete stiction intervals $I_h^j \equiv [m_j, M_j]$ for
which

$$y_n = 0 \qquad |R_{n-1}| \leqslant h\varepsilon \qquad m_j \leqslant n \leqslant M_j .$$

Although it has not been proved analytically, these discrete
stiction intervals correspond to stiction intervals of the exact
solution. Results of applying the method are described in the
next section.

4.   NUMERICAL RESULTS

Solutions have been computed for a monochromatic wave
incident on a Salter duck, shown in Figure 1, where only rotation has been considered. (Therefore forces become torques etc.)
The hydrodynamic parameters have been calculated using source-
sink methods, Count [1982], With an exciting force of the form

$$F(t) = F_0 \sin \omega_0 t \qquad (4-1)$$

the response amplitudes have been scaled by a factor

$$\left\{ \left(B - \omega_0^2 \left(I + M(\omega_0) \right) \right)^2 + \omega_0^2 N^2(\omega_0) \right\}^{\frac{1}{2}} \Big/ \omega_0 F_0 \qquad (4-2)$$

in order that an unloaded system, $\varepsilon = 0$, has a velocity amplitude
of unity. The ratio of the frictional force to the exciting
force is denoted by

$$\theta = \varepsilon/F_0 \geqslant 0 . \qquad (4-3)$$

It is clear that for $\theta = 0$ there is no stiction or discontinuous jumps in $\dot{x}$. For θ small it is also possible that
there is no stiction. If the initial conditions are that both
displacement and velocity are zero then for $\theta > 1$ no motion will

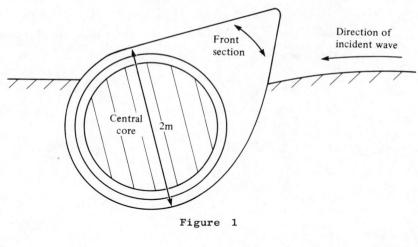

Figure 1

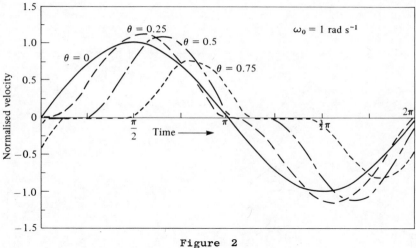

Figure 2

ever occur. Indeed if $\dot{x}(0) = 0$ and θ is sufficiently large
there will be no motion. The results of numerical calculations
are displayed in Figures $2-4$. The graphs of the normalised
velocity against time are plotted for various values of the
parameters θ and ω_0 with the remaining parameters being fixed.
Initial values were taken to be $x(0) = \dot{x}(0) = 0$ and this resulted
in steady asymptotic motion after two cycles. The periodic

solutions are given in the Figures. The graphs clearly show the
stiction intervals and the fact that as θ increases the lengths
of the intervals increase.

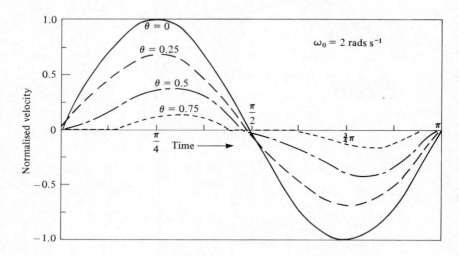

Figure 3

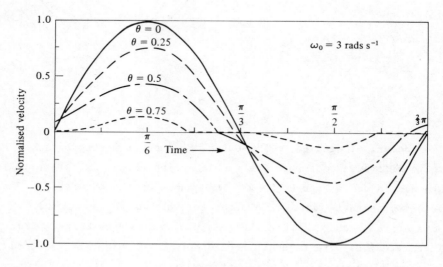

Figure 4

The numerical method has been tested by comparison with an analytic solution which can be obtained for small θ using Fourier series. The conditions under which the solution is valid are that it is periodic (due to simple monochromatic incident waves) and that continuous motion occurs with no periods of rest. Under these conditions the velocity can be assumed to vanish at only two times t_1 and t_2 during one cycle.

The steady solution can be expressed as

$$x(t) = \sum_{n=1}^{\infty} \alpha_n \cos n\omega_0 t + \beta_n \sin n\omega_0 t \qquad (4\text{-}4)$$

and $\text{sign}(\dot{x})$ can be appropriately expanded with coefficients that are dependent on t_1 and t_2. The convolution integral can similarly be evaluated in terms of the appropriate Fourier components using the analysis of Section 2.4 and the time shifted equation of motion,

$$(I + M_\infty)\ddot{x} + \varepsilon\, \text{sign}(\dot{x}) + Bx + \int_0^t G(t-\tau)\,\dot{x}(\tau)\,d\tau = F_0 \cos(\omega_0 t - \psi_1)$$

can be used to solve for the amplitudes $\{\alpha_n\}$ and $\{\beta_n\}$. After substitution and equating coefficients on either side of the equation, the harmonic amplitudes are given as

$$\alpha_n = -\frac{4\theta F_0}{\pi 2n S_n} \left[\cos(n\omega_0 t_1 - \psi_n) - \cos(n\omega_0 t_2 - \psi_n)\right]$$
$$(4\text{-}5)$$

and

$$\beta_n = \frac{\delta_{n1}}{S_1} F_0 - \frac{4\theta F_0}{2n S_n}\left[\sin(n\omega_0 t_1 - \psi_n) - \sin(n\omega_0 t_2 - \psi_n)\right]$$
$$(4\text{-}6)$$

where

$$S_n = \left[\left(B - n^2\omega_0^2(I + M(n\omega_0))\right)^2 + n^2\omega_0^2 N^2(n\omega_0)\right]^{\frac{1}{2}}, \qquad (4\text{-}7)$$

$$\psi_n = \tan^{-1}\frac{\left(B - n^2\omega_0^2(I + M(n\omega_0))\right)}{n\omega_0 N(n\omega_0)} \qquad (4\text{-}8)$$

and

$$\delta_{n1} = 1 \quad \text{if} \quad n=1 \quad \text{and} \quad 0 \text{ otherwise.}$$

If the velocity is equated to zero at t_1 and t_2, then simple solutions can be found where

$$t_2 = t_1 + \pi/\omega_0 \qquad (4\text{-}9a)$$

and

$$t_1 = \frac{1}{\omega_0} \cos^{-1} \left(\frac{4\theta}{\pi} S_1 \sum_{n=0}^{\infty} \sin \psi_{2n+1} \bigg/ S_{2n+1} \right) \qquad (4\text{-}9b)$$

and profiles of these solutions have been plotted in Figure 5 for comparison with the numerical results. It is interesting to note that whilst the convergence of the function $\varepsilon \, \text{sign}(\dot{x})$ is very slow, the actual solution converges rapidly. The reason for this is simple since the nonlinear term merely contributes harmonics to the response and their magnitude rapidly diminishes due to the inertia of the system. This is easily seen in the harmonic coefficients α_n and β_n which are proportional to $1/nS_n$ and for high harmonics this decays as $1/n^3$. This comparison between the series solution and numerical results obtained using (3-6) adds to the confidence in the discretization method given by the numerical analysis of Elliott [1985] and Elliott & McKee [1981].

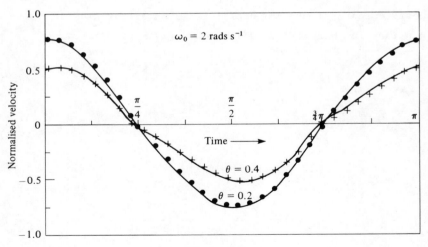

Figure   5

We now consider the efficiency of the power extraction process. When the motion is periodic the time averaged power absorbed by the device from the incident wave is given by

$$\frac{\omega_0}{2\pi} \int_{t}^{t + 2\pi/\omega_0} \epsilon \operatorname{sign}(\dot{x}) \, \dot{x} \, dt \, .$$

Assuming that an arbitrary time t_0 is chosen so that the velocity is positive between t_0 and t_1, zero between t_1 and t_2, negative in the interval t_2 to t_3 and zero for the remainder of the period, then the time averaged power absorbed is simply,

$$\frac{\omega_0 \epsilon}{2\pi} \left[x(t_1) - x(t_0) - x(t_3) + x(t_2) \right]$$

which must be compared with the average incident power per unit width,

$$\tfrac{1}{4} \omega_0 \rho \, g^2 A^2$$

where A is the incident wave amplitude.

The incident wave amplitude A is related to the forcing term F_0 by the expression, Newman [1962]

$$F_0^2 = 2 \rho g^2 A^2 \frac{N(\omega_0)}{\omega_0} \left\{ \frac{|P^+|^2}{|P^+|^2 + |P^-|^2} \right\}$$

where $P^{\pm}$ are the radiation potential amplitudes, upstream and downstream respectively, due to a unit motion at a frequency of ω_0. Noting that by symmetry

$$x(t_1) = -x(t_0) = -x(t_3) = x(t_2)$$

then the efficiency, η, reduces to

$$\eta = \frac{16 \, \Theta}{\pi} \, \omega_0 \cos \psi \, x(t_1) \left\{ \frac{|P^+|^2}{|P^+|^2 + |P^-|^2} \right\}$$

where

$$\cos \psi = \frac{\omega_0 \, N(\omega_0)}{\left[(B - \omega_0^2 (I + M(\omega_0)))^2 + \omega_0^2 \, N^2(\omega_0) \right]^{\frac{1}{2}}}$$

where $x(t_1)$ is a scaled response amplitude.

The efficiency relative to the linear optimum $|P^+|^2 / (|P^+|^2 + |P^-|^2)$, for a range of values of Θ and ω_0 is plotted in Figure 6.

When correctly tuned, the nonlinear damping mechanism is almost as good as the linear system, but in an irregular sea the discrepancy between the two systems will be increased and its significance will have to be evaluated.

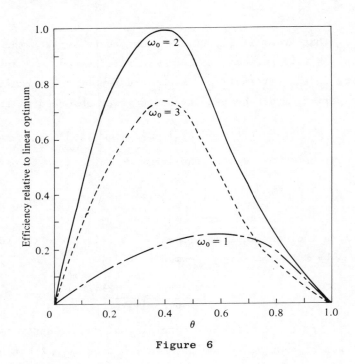

Figure 6

5   CONCLUSION

The computations which have been carried out demonstrate that the response of an excited mechanical system with Coulomb friction present can be calculated. Although the particular numerical scheme has been derived for discontinuous nonlinearities, the derived equation of motion has general application to any loading characteristic. For smooth nonlinearities multistep methods with higher order may be used.

The results presented show that the amplitude dependence of the Coulomb damping could be very important in the overall performance of such a scheme. For incident monochromatic waves, by careful choice of the damping level, the efficiency can be almost identical to the ideal linear case. If, however, the incident waves are small in comparison with the damping term then the device will appear very stiff and there will be significant periods of rest when the combined wave forces are insufficient to overcome the resistance of the load. This will be accompanied by a reduction in the efficiency of the system, unlike a linear load which adjusts the load to absorb these small waves. When irregular sea performances are calculated this inability to absorb during periods of low wave forces may be a serious drawback for any such system.

This approach to more realistic calculations on wave power devices is another step towards producing more accurate engineering design data. Detailed load characteristics are required before use can be made of the model and a great deal of parametric work will be needed. Further studies will be required on multi-degree of freedom systems.

From the numerical analysis point of view this collaboration has identified a new class of differential equations which need to be solved. Future numerical analysis research could apply the method (3-6) to other systems of differential equations with multi-valued right-hand sides and also look for more accurate methods which try to explicitly track the discontinuities.

REFERENCES

Andronov, A.A., Vitt, A.A. & Khaikin, S.E., 1966, *Theory of Oscillators*, Pergamon Press.

Count, B.M., 1977, On the hydrodynamic behaviour of ocean wave energy absorbing structures — a theoretical treatment. CEGB Laboratory Note R/M/N 960.

Count, B.M., 1982, On the physics of absorbing energy from ocean waves. Ph.D. thesis, University of Exeter.

Cummings, W.E., 1962, The impulse response function and ship motions, *Schiffstechnik*, **9**, 101-109. Reprinted as David Taylor Model Basin Report 1661.

Den Hartog, J.P., 1930, Forced vibrations with combined viscous and Coulomb damping. *Phil. Mag.* S.**7**, Vol.9, No.59.

Den Hartog, J.P., 1931, Forced vibrations with combined Coulomb and viscous friction. *Transactions of the American Society of Mechanical Engineers*, APM-53-9, 107-115.

Elliott, C.M., 1985, On the convergence of a one step method for the solution of an ordinary differential inclusion, *I.M.A.J. Numer. Anal.*, **5**. (in press).

Elliott, C.M. & McKee, S., 1981, On the numerical solution of an integro-differential equation arising from wave-power hydraulics, *B.I.T.* **21**, 318-325.

Kotik, J. & Mangulis, V., 1962, On the Kramers-Kronig relations for ship motions. *Int. Shipbuilding Progress*, **9**, No.97, 361-368.

Newman, J.N., 1962, The exciting forces on fixed bodies in waves. *Journal of Ship Research*, **6**, 10-17.

Stoker, J., 1978, *Water Waves*, Interscience.

Threfall, D.C., 1978, The inclusion of Coulomb friction in mechanisms programs with particular reference to DRAM. *Mechanism and Machine Theory*, **13**, 475-483.

7

Use of the linear functional strategy for assessing the status of plant canopies

R. S. ANDERSSEN AND D. R. JACKETT

1.   INTRODUCTION

When solving practical problems, what is done computationally must be conditioned on the data and the question which
must be answered. For example, if the problem formulation is
improperly posed, numerical methods such as regularization can
only be applied when there is a reasonable amount of data and
the variance of the observational errors they contain is small.
When the data are sparse and the variance of the errors is large,
which often occurs in practice, some alternative numerical procedure must be found which allows what little information the data
contain about the problem to be evaluated. The linear functional
strategy has proved most useful for such purposes. In essence
it reduces to simply evaluating linear functionals on the observational data which correspond to specific linear functionals
defined on the solution of the problem.

Justification for this strategy can be based on the fact
that, in applications, it is often only linear functionals defined
on the solution of the problem which must be evaluated for inference and decision making purposes.

In this chapter we illustrate the relevance of these points
for the situation when hemispherical lens photographs are used to
assess the status of plant canopies. Though we mainly consider

this specific problem, our aim in part is to illustrate a more general facet of industrial numerical analysis: *the quality of the numerical algorithm constructed to solve a given (mathematical) problem will be directly proportional to the degree to which that problem has been exploited mathematically prior to computation, and that this exploitation must be conditioned on the form of the data and the questions which must be answered.*

Knowledge about leaves in the canopies of trees is crucial in forest management, modelling canopy reflectance, studying the role of photosynthesis in plant biology, and the environmental modelling. The status and structure of plant canopies are usually assessed in terms of specific indicators, such as average foliage density, leaf area index and canopy photosynthesis, which are defined as linear functionals on the foliage angle distribution of leaves in the canopy. Among other things, the foliage angle distribution indicates how different types of trees and plants orient their leaves to either maximize or minimize the effect of sunlight at different times during the day.

For small plants, the foliage angle distribution is sometimes measured directly (cf. Lang [1973]). More generally however, measurements are made of some property of the canopy which can be related to the foliage angle distribution. In particular, measurements are made to obtain contact frequency data because, under appropriate stochastic assumptions about the spatial distribution of the foliage, the foliage angle distribution corresponds to the solution of a first kind Fredholm integral equation with the contact frequency as the known input.

In grasslands and crops, needle-like probes (point quadrats) (cf. Ross [1981], p.42) are used to record the contact frequency with the canopy for various positions of the probe. For trees, alternative techniques must be applied since their size renders infeasible any direct instrumentation method. One popular technique is to take hemispherical photographs of the canopy, and to determine from these photographs the angular averaged contact frequency with the foliage.

The determination of the foliage angle distribution from such contact frequency data is typical of the commonly occurring situation in science and technology where properties of the phenomenon of interest must be inferred from indirect measurements of that phenomenon. Consequently, the required properties are defined implicitly in terms of the data by some non-trivial mathematical formulation (a Fredholm integral equation when determining the foliage angle distribution from contact frequency data) which must be 'solved' computationally before estimates can be obtained from the data.

In addition, such mathematical formulations are improperly posed (cf. de Hoog [1980]) in the sense that their solutions are sensitive to small perturbations in the data. In fact, the degree of sensitivity, which is often measured in terms of the order of differentiation which must be applied to the data (cf. Wahba [1980]), is directly proportional to and can be taken as a measure of how indirect the observations are. For example, when the indirect observations are actually the direct, there is no sensitivity. When the indirect observations correspond to direct two-dimensional observations of a three dimensional phenomenon (for example, two-dimensional observations of 'spherical' carbon particles in steel (cf. Moran [1972])), the sensitivity corresponds to a half differentiation of the data. For the estimation of the foliage angle distribution from contact frequency data, the sensitivity is characterized by a $2\frac{1}{2}$-differentiation of the data, as we shall see below. A fractional derivative of the data $\phi(x)$ is denoted by $d^q\phi(x)/dx^q$ ($q \in \mathbb{R}$) and defined to be

$$d^q\phi(x)/dx^q = [\Gamma(-q)]^{-1} \int_0^x (x-t)^{-q-1} \phi(t)\, dt \ .$$

Even when the data are exact, the solution of such indirect measurement problems are difficult computationally. This is typical of all problems which involve (either explicitly or

implicitly) the differentiation of data. The matter is further
complicated when the data are observational. Then it is absolutely necessary to introduce some form of computational
stabilization, which effectively damps out the effect of the
differentiation of high frequency components, before acceptable
results can be obtained.

What is actually done depends heavily on the form of the
observational data and the degree of sensitivity. The higher
the degree of sensitivity, the stronger the computational stabilization which must be introduced to damp out the effect of the
higher frequency components. If the sensitivity is acute and
the observational data are poor (sparse and of high variance),
then something other than simply applying strong stabilization
must be done, since the observational data themselves may not
carry enough information about the problem to allow such a
methodology to yield a satisfactory approximation to its solution.

In such situations, the observational data to be used
carry sufficient information to allow weighted averages of the
data to be used to estimate bounded linear functionals defined
on their unknown underlying structure (because of the implicit
inherent smoothing (cf. Bloomfield [1976], p.120)). A smooth
approximation is first constructed to this underlying structure. This approximation can then be used to implement *the
linear functional strategy* of evaluating linear functionals defined on the solution of the given problem as corresponding
linear functionals defined on the exact data of the problem
(underlying structure of the observational data). As we shall
indicate below, the tacit assumption that such a correspondence
exists holds under quite general conditions. Thus, for a given
problem, the linear functional strategy can be applied once an
explicit form of the mentioned correspondence has been derived
for the functionals of interest, and a smooth approximation to
the observational data constructed.

In fact this approach has already proved quite successful

in extracting from observational data useful information about
the solution of the problem in terms of appropriate linear functions (cf. Golberg [1979] and Anderssen [1980]).

Justification for the linear functional strategy can be
based on the fact that, in applications, it is often only linear
functionals defined on the solution of the problem which are
required for inference and decision making purposes, and not the
solution itself. For example, in the study of plant canopies,
average foliage density, leaf area index and canopy photosynthesis as well as the standard moments are the key measures which
correspond to functionals defined on the foliage angle distribution (cf. Ross [1981] and Anderssen, Jackett & Jupp [1984]).

2.   THE PROBLEM AND MATHEMATICAL MODEL

In the study of the structure and status of plant canopies,
the aim is to use indirect measurements made on the canopy
to obtain useful information about it. As mentioned briefly in
the introduction, various indirect measurement procedures can be
used. We limit attention to the use of hemispherical lens
cameras.

The starting point for the construction of a mathematical
model can therefore be described in the following way. A hemispherical lens camera is placed on the (horizontal) ground below
the canopy of interest and photographs of the type illustrated
in Figure 1 taken. The spider's-web grid of Figure 2 is then
centred on the photograph and used to evaluate the average gap
$P_0(\beta)$ in each annulus of the grid as a function of the radius β
of the mid-point of the annulus. This is done by averaging the
proportion of gap in each of the sectors forming the annulus.
This averaged gap $P_0(\beta)$ can then be reinterpreted as the azimuthally averaged gap $P_0(\beta)$ as a function of the elevation angle
β of the azimuthal annulus above the ground. The rationale used
by Anderson [1964 , 1971] for the construction of the grid shown
in Figure 2 is that each division of the grid contributes equal

radiance from a Standard Overcast Sky (Walsh [1961]) onto the
horizontal surface.

Figure 1
A hemispherical ('fish-eye')
lens camera photograph of a
plant canopy

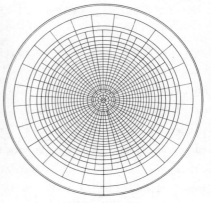

Figure 2
Radial spider-web grid used
to evaluate the average gap
in each annulus of the grid

Now information about the structure of the canopy must be
based on some quantitatively defineable property of the leaves.
Since, from the point of view of photosynthesis, the surfaces
rather than the edges of the leaves are the important biological
feature, the leaves are usually modelled (to a first approximation) as planar surfaces. A planar surface, however, is defined
by two characteristics: its area and its normal. We therefore
need a property which incorporates both these concepts. Because
mature leaves are more or less of the same size, it is natural
to conclude that the normals are more important than the areas.
Further reflection confirms this conclusion. Because of the
structure of trees, the normals to the leaves exhibit greater
variability than their areas. In addition, in a photosynthesis
calculation, the angular differences between the direction of
the sunlight and the normals to the leaves are more important
than their individual areas (though clearly the areas cannot be
ignored).

For these reasons, the leaves are modelled as the percentage of the total area which has a given direction, with the

individuality of the leaves being incorporated into this single area.

Since the normals of the leaves do not follow a clearly defined deterministic pattern, their angle distribution must be modelled stochastically. The probability density function model in popular use is the azimuthally averaged foliage angle density function $g(\alpha)$, where $g(\alpha)\,d\alpha$ defines the contribution to the foliage area of leaves inclined to the horizontal at angles between α and $\alpha + d\alpha$. This is based on the assumption that the azimuth angle distribution of the foliage normals is uniform. Clearly, this model replaces the locational variation with a grouped variation of the type alluded to above.

It turns out that $g(\alpha)$ is an excellent way in which to quantitatively model the leaves in a canopy. For example, many important properties of a canopy, such as leaf area index and photosynthesis indicators, can be defined as linear functionals on $g(\alpha)$. In addition, models can be derived which relate $g(\alpha)$ and $P_0(\beta)$ mathematically.

The latter is achieved by making appropriate assumptions about $P_0(\alpha)$ and $g(\alpha)$. In fact, if it assumed that the leaves are of equal area with centroids randomly distributed by a spatially stationary (Poisson) distribution, it can be shown that (cf. Nilson [1971] and Smith et $al.$ [1977])

$$P_0(\beta) = \exp\left(-f(\beta)/\sin\beta\right), \qquad (2\text{-}1)$$

or equivalently

$$f(\beta) = -\sin\beta\,\ln(P_0(\beta)) \qquad (2\text{-}2)$$

where $f(\beta)$ denotes the azimuthal averaged mean canopy projection in the elevation direction β. This projection is the point of common reference between the indirect measurements and the required $g(\alpha)$. On the one hand, it can be shown (cf. Reeve's Appendix in Warren Wilson [1960], Philip [1965], and Smith et $al.$ [1977]) that

$$f(\beta) = \int_{0}^{\pi/2} k(\alpha,\beta)\,g(\alpha)\,d\alpha, \qquad 0 \leqslant \alpha \leqslant \pi/2, \quad (2\text{-}3)$$

where

$$k(\alpha, \beta) = \begin{cases} \cos \alpha \sin \beta & \alpha \leqslant \beta, \\ \cos \alpha \sin \beta \left[1 + 2/\pi (\tan \theta - \theta)\right] & \alpha \geqslant \beta, \end{cases} \qquad (2\text{-}4)$$

with

$$\theta = \cos^{-1}(\tan \beta / \tan \alpha), \qquad 0 \leqslant \theta \leqslant \pi/2. \qquad (2\text{-}5)$$

On the other, $f(\beta)$ can be determined from various types of indirect measurements (cf. Jupp *et al.* [1980]).

The integral equation $(2\text{-}3) - (2\text{-}5)$ has been studied mathematically in some detail. Under the conditions that f is such that f'' exists and is absolutely continuous on $[0, \pi/2]$ and $f'(0) = f'(\pi/2) = 0$, Miller [1963] derived the following inversion formula

$$g(\alpha) = \tan \alpha \ \sec^3\alpha \int_{\alpha}^{\pi/2} \frac{3 \cos^2\tau \sin \tau (f(\tau) + f''(\tau))}{(\tan^2\tau - \tan^2\alpha)^{\frac{1}{2}}} -$$

$$- \frac{\cos^3\tau \ (f'(\tau) + f'''(\tau))}{(\tan^2\tau - \tan^2\alpha)^{\frac{1}{2}}} \, d\tau, \qquad (2\text{-}6)$$

where $f'(\tau) = df(\tau)/d\tau$. Subsequently, after recalling that the Abel integral equation has two inversion formulae (cf. Kowaleski [1930]), Anderssen & Jackett [1984] derived the alternative form

$$g(\alpha) = -\sec \alpha \ \frac{d}{d\alpha} \left\{ \int_{\alpha}^{\pi/2} \frac{\sin \beta (f(\beta) + f''(\beta))}{(\tan^2\beta - \tan^2\alpha)^{\frac{1}{2}}} \, d\beta \right\}. \qquad (2\text{-}7)$$

A number of authors have observed that the integral equation $(2\text{-}3) - (2\text{-}5)$ is improperly posed in the sense that its solution is sensitive to small perturbations in the contact frequency $f(\beta)$. Smith, Oliver & Berry's [1977, Appendix] observation is based on the general characterization of $(2\text{-}3)$ as a first kind Fredholm integral equation and the fact that the discretization of such equations yield matrix equations which are poorly conditioned (i.e. the linear independence between columns of the matrices is weak). The observation of Anderssen & Jackett is based on the existence of a third derivative of $f(\beta)$ under the integral sign in the inversion formula $(2\text{-}6)$. By analogy with the Abel integral equation (cf. Sneddon [1966]), the indefinite singular integral equation in (2-6) (and (2-7)) corresponds to a $'-\frac{1}{2}'$ differentiation. Thus, the evaluation of either the inversion formula (2-6) or (2-7) involves a $2\frac{1}{2}$-differentiation of the contact frequency $f(\beta)$.

3.   THE DATA AND COMPUTATIONAL CONSEQUENCES

Together, equations (2-1) — (2-5) define the relationship between the indirect photographic measurements of $P_0(\beta)$ and the phenomenon of interest $g(\alpha)$. However, this does not solve the problem, but only yields a foundation on which the study of plant canopies can be built. In reality, the practitioner not only requires a recipe for solving (2-3) — (2-5), but wants the value of some key indicators such as the leaf area index F and moments $M_p(g)$

$$F = \int_0^{\pi/2} g(\alpha)\, d\alpha\ , \qquad M_p(g) = \int_0^{\pi/2} \alpha^p\, g(\alpha)\, d\alpha\ , \qquad (3\text{-}1)$$

given that the only information about $f(\beta)$ is the data

$$d_i = f(\beta_i) + \varepsilon_i\ , \quad 0 \leqslant \beta_1 < \beta_2 < \ldots < \beta_n \leqslant \pi/2\ , \quad (3\text{-}2)$$

where the ε_i denote observational (random) errors, and given that n is small and the variance of the ε_i is large.

Note: Because it is a simple matter to use (2-2) to estimate $f(\beta)$ from the measured values of $P_0(\beta)$, the need to solve (2-1) — (2-5) is often simplified to solving (2-3) — (2-5) for the derived data d_i, $i = 1, 2, \ldots, n$, of (3-2).

If no effort is first made to smooth the measured values of $P_0(\beta)$ before applying (2-2), the derived data usually takes a form similar to that shown in Figure 3. It is obviously reasonably sparse. In addition, a comparison with the synthetic data of Figure 4 illustrates that their errors have a high variance.

For this reason, the derived data (3-2) cannot be used as it stands for the solution of (2-3) — (2-5). It is clear that, because of the differentiations involved, the finite difference versions of either (2-6) or (2-7) would yield completely

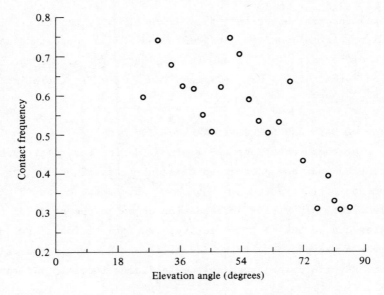

Figure 3. Contact frequency data obtained
from a hemispherical lens camera photograph

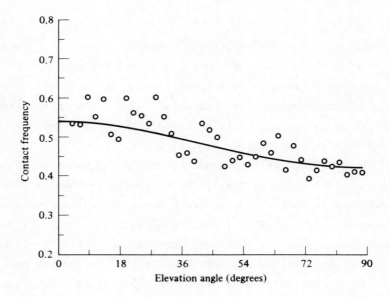

Figure 4. Synthetic data constructed to illustrate
the high variance nature of contact frequency data

unrealistic results if applied to this data. In addition, matrix regularization performs poorly. Because the regularization parameter must be made large to ensure a smooth result and to damp out the effect of the implicit $2\frac{1}{2}$-differentiation of the higher frequencies, the approximation it generates relates more to the null space structure of the regularizing functional being used than to the specific solution of the integral equation (2-3) — (2-5) which corresponds to the underlying structure of the data.

This is a direct consequence of theoretical results about the dependence of the regularization solution on the regularization parameter (cf. Lukas [1980], and Schoenberg [1964]). They show that, as this parameter tends to zero, the solution tends to a certain solution which satisfies the data exactly; and, as the parameter tends to infinity, the solution tends to a smooth function which corresponds to a smooth approximation of the data which lies in the null space of the functional being used in the regularization. A simple illustration of these facts can be found in most discussions of the use of splines to smooth data using the corresponding regularization formulation (cf. Schoenberg [1964]).

The next step is to fit some smooth (parametric) representation to the underlying trend in either the measured $P_0(\beta)$ or in the derived data (3-2), and thereby obtain a smooth approximation $\hat{f}(\beta)$ to $f(\beta)$.

Both the inversion formulae (2-6) and (2-7), and regularization, were applied to $\hat{f}(\beta)$. With care, better results were obtained than when the unsmoothed data was used; but they depended heavily on the parametric form used to define $\hat{f}(\beta)$. Because of the limited number of data points defining $P_0(\beta)$ and $f(\beta)$, there is great scope for choosing the parameterization of $\hat{f}(\beta)$. Numerical experimentation with synthetic data, for which the solution of (2-3) — (2-5) is known explicitly for a given $f(\beta)$, was used to test various possibilities.

When either the inversion formula (2-6) or (2-7) was applied to $\hat{f}(\beta)$, it was difficult to find the compromise between the following two extreme situations:

(i) a simple (low parameterization) model for $\hat{f}(\beta)$ (e.g. low degree polynomial) which yields smooth higher derivatives but at the expense of obtaining poor approximations to the correct derivatives;

(ii) a flexible (high parameterization) model for $\hat{f}(\beta)$ (e.g. a smoothing spline) which has the potential to yield good approximations to the higher derivatives, but at the expense of being too sensitive to the $2\frac{1}{2}$-differentiation.

The application of regularization encountered similar difficulties. The use of a simple model for $\hat{f}(\beta)$ gave poor but smooth approximations to the correct answer, while the use of a flexible model needed a very large regularization parameter before it generated a smooth approximation. Unfortunately, because of the large value of the regularization parameter, this approximation tended to oversmooth the correct answer.

If a large amount of accurate data had been available, some of these difficulties would have been circumvented by either using stabilized differentiation to assist with the evaluation of the inversion formulae or by applying matrix regularization. Nevertheless, the inherent $2\frac{1}{2}$-differentiation would not have made this an easy task to accomplish.

At this stage, it becomes clear that for the study of plant canopies, somthing different must be done with the contact frequency (or average gap) data if the little information they contain is to be 'painlessly' extracted and successfully interpreted. It is one thing to say something different must be done, it is another to decide how.

Thus, in some way or other, a process of lateral thinking about the problem must be initiated. One way is to appeal to past or established experience in terms of the general guidelines

it yields about the art and nature of industrial problem solving.
Among other things, such sources (cf. Anderssen & de Hoog [1981,
Introduction]) invariably stress the need to

> "ensure right from the start that the question being
> answered is the question which should have been
> asked. The form of the mathematics used depends
> heavily on the question which must be answered."

With this point in mind, one recalls that, when defining
the study of plant canopies in Section 2, it was stressed that
the aim was 'to use indirect measurements made on the canopy to
obtain useful information about it'. Up to now, it has been
tacitly assumed that the useful information is $g(\alpha)$. In many
situations, however, the useful information consists of one or
two key indicators (e.g. leaf area index and the moments of $g(\alpha)$)
which will be used for decision making purposes.

Thus, from an industrial problem solving point of view,
the study of plant canopies is reduced to answering the question:
*when and how can the key indicators be computed directly from the
given contact frequency data* (3-2) *without first computing the
foliage angle distribution* $g(\alpha)$?

When the indicators correspond to linear functionals defined on $g(\alpha)$, this can be achieved by applying the linear
functional strategy.

It is important to appreciate that the mentioned lateral
thinking cannot start the moment the problem is first considered.
There is a need to 'understand' the problem before the process of
lateral thinking can be initiated. In part, this is the motiva-
tion and justification for the format of the above discussion.

4.   THE APPLICATION OF THE LINEAR FUNCTIONAL STRATEGY

As mentioned in Section 1, *the linear functional strategy*
consists of evaluating linear functionals on the solution of some
problem as corresponding linear functionals defined on the exact
data of that problem. Thus for a general problem:

$$Kg = f \qquad (4\text{-}1)$$

with K a linear operator, it can be defined as follows: *for a given function* $\theta(\alpha)$ *which defines the linear functional*

$$L_\theta(g) = \int_0^{\pi/2} \theta(\alpha)\, g(\alpha)\, d\alpha \qquad (4\text{-}2)$$

on the solution of (4-1), *determine the function* $\phi(\beta)$ (assuming it exists) *for which*

$$L_\theta(g) = \int_0^{\pi/2} \phi(\beta)\, f(\beta)\, d\beta \;. \qquad (4\text{-}3)$$

Clearly, because the form of (4-3) can involve boundary terms (e.g. $f(0)$, $f(\pi/2)$, $f'(0)$,...), $\phi(\beta)$ will on occasions have a distributional interpretation.

Formally, if the inverse operator K^{-1} exists, the transformation can be accomplished using

$$L_\theta(g) = \int_0^{\pi/2} \theta(\alpha)\, (K^{-1} f)(\alpha)\, d\alpha \qquad (4\text{-}4)$$

though a meaningful interpretation as the right hand side of (4-3) will only be possible when appropriate restrictions are placed on θ relative to (4-1).

Under the assumption that K is a densely defined mapping from a Hilbert space $\mathcal{H}$ into itself with inner product

$$(u,v) = \int_0^{\pi/2} uv\; dx \;, \qquad u,v \in \mathcal{H}, \qquad (4\text{-}5)$$

a rigorous characterization of the transformation from $\theta(\alpha)$ to $\phi(\beta)$ is given by

$$K^* \phi = \theta \qquad (4\text{-}6)$$

under the restriction that $\theta \in \mathcal{R}(K^*)$ (i.e. the range of the adjoint operator K^*) (cf. Golberg [1979]). Using the inner product notation, this is easily verified on observing that

$$L_\theta(g) = (\theta, g) = (K^* \phi, g) = (\phi, Kg) = (\phi, f) \;. \qquad (4\text{-}7)$$

As mentioned above, from a numerical point of view, the

appeal of this result is the ease with which it copes with
the common practical situation where the only information available about f is observational data of the form (3-2), since the
estimation of $L_\theta(g)$ is reduced to evaluating (ϕ, f) using only
the data (3-2).

The solution of $Kg = f$ is replaced by the need to solve
$K^*\phi = \theta$. But, when K^* and θ are such that ϕ can be determined
analytically (e.g. above, and the Abel integral equation case
(Anderssen [1980])), the overall computational endeavour is
greatly simplified. Even in situations where it is in fact
necessary to solve $K^*\phi = \theta$ numerically to determine an approximation to ϕ, the position is far superior to that defined by
(4-1) and (4-2), since θ is known analytically.

However, it is one thing to have such theoretical charac-
terizations, it is another to find a way to implement them for a
specific problem. This is another aspect of industrial problem
solving: the need to work with the specific structure under
examination and nothing else. For the linear functional strategy,
what is actually done depends heavily on the form of θ.

For the foliage angle distribution problem (2-3) — (2-5),
one can in fact utilize the formal representation (4-4) through
the use of the inversion formulae. Thus, the need for the second
inversion formula becomes apparent as it allows an immediate
integration by parts (cf. the corresponding results for Abel's
equation). Using it, Anderssen & Jackett [1984] have proved
that

$$L_\theta(g) = \theta(0) f(\pi/2) + \int_0^{\pi/2} \psi(\beta) \sin\beta \left[f(\beta) + f''(\beta)\right] d\beta ,$$

(4-8)

with

$$\psi(\beta) = \int_0^\beta \frac{d}{d\alpha}(\sec\alpha \, \theta(\alpha)) \frac{\cos\alpha \, \cos\beta}{\sqrt{\sin^2\beta - \sin^2\alpha}} d\alpha \qquad (4\text{-}9)$$

under the restriction that $f''(\beta)$ exists and is absolutely continuous on $[0, \pi/2]$, $f'(0) = f'(\pi/2) = 0$, and $\theta'(\alpha)$ exists and
is continuous on $[0, \pi/2]$.

This result clearly illustrates the point made above about the implementation for a specific problem. The $\psi(\beta)$ of (4-9) and hence the right hand side of (4-8) will only assume a simple form for special choices of $\theta(\alpha)$. For example (cf. Anderssen & Jackett [1984]), when $\theta(\alpha) = 1$, $\sin\alpha$, and $\cos\alpha$, $L_\theta(g)$ reduces to

$$2 \int_0^{\pi/2} \cos\beta \; f(\beta) \; d\beta \; , \quad \pi f(0)/2 \quad \text{and} \quad f(\pi/2)$$

respectively. When $\theta(\alpha) = \alpha^p$, $p = 1, 2, \ldots$, the explicit forms for $\psi(\beta)$ can be obtained by integration by parts, but they are too complex to be of practical use computationally. When

$$\theta(\alpha) = \chi[a,b] = \begin{cases} 1 & \text{if } \alpha \in [a,b], \\ 0 & \text{if } \alpha \notin [a,b]; \end{cases} \qquad (4-10)$$

the transformation (4-9) cannot be applied. In this case, Anderssen & Jackett [1984] and Jackett & Anderssen [1984], have used the first inversion formula to derive a number of independent relationships for $L_{\chi[a,b]}(g)$.

From the point of view of the present discussion, it is of interest to note that, in the early work on estimating the leaf area index, experimental procedures (cf. Warren Wilson [1963]) were used to represent it as a linear combination of measured values of the contact frequency $f(\beta)$; viz.

$$F \doteq w_1 f(\beta_1) + w_2 f(\beta_2) + \ldots + w_m f(\beta_m) \qquad (4-11)$$

where the w_i and the β_i, $i = 1, 2, \ldots, m$, denote predetermined weights and elevation angles. Subsequently, Miller [1967] showed, using the first inversion formula (2-6), that

$$F = 2 \int_0^{\pi/2} \cos\beta \; f(\beta) \; d\beta , \qquad (4-12)$$

and thereby formalized mathematically Warren Wilson's 'experimental' approach: any (non-derivative) quadrature formula for (4-12) defines a corresponding formula for the right hand side of (4-11).

The result (4-8) and (4-9) has two weaknesses. Firstly,

the functional involves a second differentiation of f, and therefore where possible this differentiation should be transferred to ψ. Secondly, the evaluation of ψ involves a (square root) singularity which, for most choices of $\theta(\alpha)$ produce elliptic integrals. Since these integrals must in general be evaluated by numerical means, thereby introducing errors into the specification of ψ itself, this negates any advantages with moving the differentiation from f to ψ.

What is required is an alternative approach which will yield a computationally useful procedure for difficult $\theta(\alpha)$. Jackett & Anderssen [1984] have derived such a procedure using a generalization of the leaf area index

$$I(f;\alpha) = \int_{\alpha}^{\pi/2} \frac{\partial \sqrt{\sin^2\beta - \sin^2\alpha}}{\partial \beta} f(\beta) \, d\beta . \qquad (4\text{-}13)$$

Clearly, $2I(f;0) = F$, the leaf area index as defined by (3-1).

Despite the elliptic nature of $I(f;\alpha)$, it can be computed very accurately and efficiently using product integration. Indeed, on an even grid of $(n+1)$ points on the interval $[\alpha, \pi/2]$, the strategy is to approximate $f(\beta)$ by a piecewise constant so that the derivative under the integral sign can be exploited analytically. In fact, the product mid-point integration scheme yields the following approximation:

$$\hat{I}(f;\alpha) = \sum_{i=0}^{n-1} \left\{ f(\bar{\beta}_i) \sqrt{\sin^2\beta_{i+1} - \sin^2\alpha} - f(\bar{\beta}_i) \sqrt{\sin^2\beta_i - \sin^2\alpha} \right\} \qquad (4\text{-}14)$$

where $\bar{\beta}_i = (\beta_i + \beta_{i+1})/2$, $\beta_i = \alpha + ih$, $i = 0, 1, \ldots, n$ and $h = (\pi - 2\alpha)/2n$. Clearly, its advantage is that it does not involve any differentiation of the contact frequency f.

Because $I(f;\alpha)$ can be computed accurately and efficiently using (4-14), it can be used to simplify the evaluation of a number of functionals. For example, using (4-10), Jackett & Anderssen [1984] have derived the following simple representation for the segmented foliage density in terms of $I(f;\alpha)$

$$\int_{\alpha}^{\pi/2} g(\alpha)\,d\alpha = 2I(f;\alpha) - \cot\alpha I'(f;\alpha) + I''(f;\alpha)\,, \qquad (4\text{--}15)$$

where the differentiation is with respect to α.

Here, the differentiation on f in the inversion formula has been transferred to a (lower order) differentiation of $I(f;\alpha)$. Because $I(f;\alpha)$ corresponds to a smoothing of the contact frequency data, the right hand side of (4-15) has obvious computational advantages over the explicit evaluation of its left hand side.

More importantly, however, is the role played by $I(f;\alpha)$ in the following computational formula for $L_\theta(g)$

$$\int_0^{\pi/2} \theta(\alpha)\,g(\alpha)\,d\alpha = -\int_0^{\pi/2} \left\{ 2\theta(\alpha) + \cot\alpha\,\theta'(\alpha) + \theta''(\alpha) \right\} I'(f;\alpha)\,d\alpha$$
$$-\theta'(\pi/2)\,f(\pi/2) \qquad (4\text{--}16)$$

derived by Jackett & Anderssen [1984], under the rather weak restriction on $\theta(\alpha)$ that $\theta'(\alpha)$ exists and is absolutely continuous. The differentiation in (4-16) poses no difficulty. Because $\theta(\alpha)$ is known analytically

$$\Psi(\alpha) = -\left\{ 2\theta(\alpha) + \cot\alpha\,\theta'(\alpha) + \theta''(\alpha) \right\} \qquad (4\text{--}17)$$

can be determined analytically. Product integration is now used to replace $\Psi(\alpha)$ by a piecewise approximation so that the differentiation of $I(f;\alpha)$ can be avoided and thereby allow the direct use of the estimates $\hat{I}(f;\alpha)$ of (4-14). Again using the product mid-point integration scheme on an even grid of $(n+1)$ points on $[0,\pi/2]$, the evaluation of

$$I_\Psi(f) = \int_0^{\pi/2} \Psi(\alpha)\,I'(f;\alpha)\,d\alpha \qquad (4\text{--}18)$$

is reduced to

$$\hat{I}_\Psi(f) = \sum_{i=0}^{n-1} \Psi(\bar{\beta}_i)\left\{ I(f;\beta_{i+1}) - I(f;\beta_i) \right\}. \qquad (4\text{--}19)$$

Using synthetic data for representative plant canopies, Anderssen *et al.* [1984] have shown that the above linear functional strategy yields more accurate results more efficiently than any direct numerical procedure based on the solution of (2-3) — (2-5).

It is clear that these results yield a general framework for the application of the linear functional strategy to the foliage angle distribution problem, and that this has only been possible through the manipulation of the underlying mathematical structure of the problem. In addition, the need to develop this general framework arose out of the fact that, in the study of plant canopies, it is not simply a recipe for solving (2-3) — (2-5) which is required, but the values of some key indicators $L_\theta(g)$, given that the only information about $f(\beta)$ were the data (3-2).

5.   CONCLUSION

In this case study, we have shown how (linear) foliage angle functionals can be computed directly from available observational data as contact frequency functionals, and indicated that accurate estimates are obtained even when the data are sparse and quite noisy. The advantage and efficiency of exploiting mathematically the structure of the problem to achieve such results have been discussed. In particular, it is explained in Section 4 how simple expressions can be derived for the contact frequency functionals corresponding to the foliage angle functionals of interest.

The availability of such formulae is assisting greatly with projecting the utility of the idea that, for decision making purposes concerned with the structure and status of plant and crop canopies, the botanist and agronomist should identify and work with appropriate indicators such as specific foliage angle functionals. In fact, previously, the full potential of such an approach had not been exploited except for the use of the leaf area index, because of a lack of simple and accurate formulae which allowed desired indicators to be determined directly from the available contact frequency data.

Acknowledgement. Both authors wish to thank Drs M.C. Anderson and D.L.B. Jupp for their assistance during the work on the foliage ɩngle problem. Dr Anderson has supplied data and unpublished photographs of canopies, and has given permission for the publication of the material in Figures 1 and 3. Dr Jupp introduced us to the problem and has supplied valuable background information during our investigations.

REFERENCES

Anderson, M.C. 1964, Studies of the woodland climate. I. The photographic computation of light conditions, *J. Ecol.* **52**, 27-41.

Anderson, M.C. 1971, Radiation and crop structure, in *Plant Photosynthetic Production/Manual of Methods* (Eds J. Catsky and P.G. Jarvis), Junk, The Hague, 412-466.

Anderssen, R.S. 1980, On the use of linear functionals for Abel-type integral equations in applications, in *The Application and Numerical Solution of Integral Equations* (Eds R.S. Anderssen, F.R. de Hoog & M.A. Lukas), Sijthoff and Noordhoff, Alpen aan den Rijn.

Anderssen, R.S. & de Hoog, F.R. 1982, *The Application of Mathematics in Industry* Martinus Nijhoff, The Hague.

Anderssen, R.S. & Jackett, D.R. 1984, Linear functionals of foliage angle density, *J. Aust. Math. Soc., Series B* **25**, 431-442.

Anderssen, R.S., Jackett, D.R. & Jupp, D.L.B. 1984, Linear functionals of the foliage angle distribution as tools to study the structure of plant canpies, *Aust. J. Bot.* **32**, 147-156.

Bloomfield, P. 1976, *Fourier Analysis of Time Series*, John Wiley and Sons, New York.

de Hoog, F.R. 1980, Review of Fredholm integral equations of the first kind, in *The Application and Numerical Solution of Integral Equations* (Eds R.S. Anderssen, F.R. de Hoog & M.A. Lukas), Sijthoff and Noordhoff, Alpen aan den Rijn.

Golberg, M.A. 1979, A method of adjoints for solving some ill-
posed equations of the first kind, *Applied Math. and Comp.*
5, 123-130.

Jackett, D.R. & Anderssen, R.S. 1984, Computing the foliage
angle distribution from contact frequency data, in *Computational Techniques and Applications*, CTAC-83 (Eds J. Noye
& C. Fletcher), North Holland, Amsterdam.

Jupp, D.L.B., Anderson, M.C., Adomeit, G.M. & Witts, S.J. 1980,
PISCES — A computer program for analyzing hemispherical
canopy photographs, Tech. Memo 80/23, CSIRO Division of
Land Use Research.

Kowalewski, G. 1930, *Integralgleichungen*, Walter de Gruyter &
Co., Berlin and Leipzig.

Lang, A.R.G. 1973, Leaf orientation of a cotton plant, *Agric.
Meteorol.* 11, 37-51.

Lukas, M.A. 1980, Regularization, in *The Application and
Numerical Solution of Integral Equations* (Eds R.S. Anderssen,
F.R. de Hoog & M.A. Lukas), Sijthoff and Noordhoff, Alpen aan
den Rijn.

Miller, J.B. 1963, An integral equation from Phytology, *J. Aust.
Math. Soc.* 4, 397-402.

Miller, J.B. 1967, A formula for average foliage density, *Aust.
J. Bot.* 15, 141-144.

Moran, P.A.P. 1972, The probabilistic basis of stereology,
Special Supplement to Adv. Appl. Prob. 4, 69-91.

Nilson, T. 1971, A theoretical analysis of gaps in plant stands,
Agric. Meteorol. 8, 25-38.

Philip, J.R. 1965, The distribution of foliage density with
foliage angle estimated from inclined point quadrat observations, *Aust. J. Bot.* 13, 357-366.

Ross, J. 1981, *The Radiation Regime and Architecture of Plant
Stands*, Junk, The Hague.

Schoenberg, I.J. 1964, Spline functions and the problem of
graduation, *Proc. Nat. Acad. Sci.* 52, 947-950.

Smith, J.A., Oliver, R.E. & Berry, J.K. 1977, A comparison of
two photographic techniques for estimating foliage angle distribution, *Aust. J. Bot.* 25, 545-553.

Sneddon, I. 1966, *Mixed Boundary Value Problems in Potential
Theory*, North-Holland, Amsterdam.

Wahba, G. 1980, Ill-posed problems: numerical and statistical
methods for mildly, moderately and severely ill-posed problems with noisy data, Tech. Report 595, University of
Wisconsin, Department of Statistics.

Walsh, J.W.T. 1961, *The Science of Daylight*, MacDonald, London.

Warren Wilson, J. 1960, Inclined point quadrats, with Appendix by J.E. Reeve, *The New Phytologist* **59**, 1-8.

Warren Wilson, J. 1963, Estimation of foliage denseness & foliage angle by inclined point quadrats, *Aust. J. Bot.* **11**, 95-105.

8

The ageing of stainless steel

or how to reduce a stiff system of diffusion equations to an integral equation

J. NORBURY

1.   INTRODUCTION

The problem of the ageing of stainless steel is of concern to the Central Electricity Generating Board because of the need to design and assess the performance of many components in present day nuclear reactors. The type of stainless steel used consists of an iron matrix with 15−20% by weight of chrome and 0.1% by weight of carbon uniformly distributed throughout the grain structure.

The chrome protects against corrosion while the carbon provides mechanical strength. During the high temperature (say 600°C) use of this stainless steel in the reactor it is observed that carbon and chrome slowly precipitate out, mainly as chromium carbides. The loss of chrome and carbon affect both the ability of the steel to withstand corrosive attack and its mechanical properties.

The chrome and carbon atoms are uniformly mixed throughout the steel at temperatures of around 1100°C in the annealing part of the manufacture of the steel. At these temperatures the carbon and chrome atoms are essentially dissolved in saturated solution, and are free to move throughout the steel in a matter of hours. At the lower operating temperatures of 500−800°C these atoms are now much less mobile, but they are also highly

supersaturated in the steel, and precipitate out by reaction at
certain sites either on the grain boundaries or at defects.

At the temperatures we consider, the carbon atoms are still
very mobile: a carbon atom can diffuse (migrate) across a grain
to a reaction site in a matter of days. However the chrome atoms
are highly immobile in the grains, taking about a thousand years
to effect a similar journey. Since both carbon and chrome must
be present at the reaction sites in the ratio of about one to
three in order to form the chromium carbide ($M_{23}C_6$ where M is
a mixture of mostly chrome and iron metallic elements), the problem is 'how does this occur?' In more detail, how can the
highly mobile carbon interact with the highly immobile chrome on
an intermediate time-scale of tens of years so as to seriously
affect the properties of the stainless steel? Our mathematical
model of this process leads to a problem involving two coupled
diffusion equations for the two species, where the diffusivity
of one species is many orders of magnitude greater than the
diffusivity of the other species.

This enormous disparity in diffusivities (and in typical
migration time-scales) leads to rapid equilibrium for the concentration profile of the carbon after several days, while a steep
boundary layer structure near the grain boundary in the metal
appears in the chrome concentration. Thus the time-scale
disparity leads to a space scale disparity, and this spatial
disparity remains for hundreds of years.

During this time the whole carbon concentration gradually
ebbs away, while the chrome is leached from around the grain
boundaries. (In fact, experiments with acid attack can actually
dislodge the grains of metal, since the loss of chrome around
the boundary leaves the iron atoms in the boundary vulnerable.)
Over the time-scale of practical interest, say five to twenty
years, the carbon levels throughout the grain can drop by 50%,
while a similar drop in the chrome concentration can penetrate
about one micron into the grain (whose typical dimension we take

as about 100 microns).

Thus it is clear that a direct numerical attack on problems with widely differing diffusivities (stiff heat equations) will be difficult. As described in Markham [1980] the boundary layer structure of the slowly diffusing species has to be resolved appropriately, with either a finite difference or a finite element mesh. The species with large diffusion has to be followed in near equilibrium for thousands of its natural time steps. The grain boundaries are not easy to simulate numerically, and the time dependent problem is fully three space dimensional. The initial conditions imply infinite derivatives initially. Finally there is a nonlinear linkage between the species, which means some sort of iteration is needed at each time step. Markham [1980] took a simplified spherical geometry, exploited the boundary layer structure and could integrate forward about one thousandth of the time-scale required.

To proceed to the time-scale of interest requires a new approach, Skyrme brought the original problem to the inaugural University Consortium for Industrial Numerical Analysis meeting in Oxford in January, 1980, and the author suggested an analytical approach to reduce the coupled partial differential equation problem to one involving only a single nonlinear Volterra integral equation. An outline of this approach, together with details of the original model, and comparisons between theory and experiments are given in Skyrme & Norbury [1980].

The ideas behind this approach are as follows. We first locate the chrome boundary layer and evaluate the flux of chrome through the layer in terms of the difference in chrome concentration across the boundary layer. Next we relate the slow decay of the carbon concentration to the flux of carbon at the grain boundary where the carbon disappears. The carbon and chrome fluxes must be in the ratio that makes the metallic chromium carbide of fairly uniform composition, say two to three. Thus,

so far, we have related the decay of the average carbon concentration to the chrome concentration at the precipitation site.
Finally we need to use some precipitation chemistry, which
relates the carbon and chrome concentrations at the reaction
sites by a nonlinear equilibrium law, to obtain a single
integral equation for the carbon concentration over the required
time-scale.

Thus the original difficult problem of coupled partial
differential equations with fairly singular behaviour has been
reduced to a much simpler integral equation for a regular
function of one variable. The moral of this case study would
appear to be that the numerical solution of singular problems
often benefits from a preliminary simplifying mathematical analysis in order to find a more regular problem.

2.   THE PHYSICAL MODEL

Considerable experimental work has led to a basic understanding of carbide formation in stainless steels. Theoretical
models of precipitate growth by diffusion are discussed in
Kircaldy [1958] and Coates [1972-3], for instance. However the
effect of finite geometries on the carbon concentration was not
considered until Goldstein & Randich [1977]. Their numerical
scheme only gives reasonable results for comparable diffusivities. The work of Snyder *et al.* [1973] is based on the assumption
that carbide precipitate growth is controlled by the diffusion
of carbon — this is the opposite of what we find.

The model that we adopt follows that of Strawström &
Hillert [1969] and Tedmon *et al.* [1971]. Although highly simplified, the model produces results in excellent agreement with
experiments, as discussed in Skyrme & Norbury [1980]. We assume
that the carbide precipitation occurs at sites located on grain
boundaries (the case of carbide formation at defects within the
grain follows similarly), the precipitate being essentially
$M_{23}C_6$, where twenty-three metal atoms (about eighteen of which

are chrome) combine with six carbon atoms. The carbon and chrome
atoms diffuse through the grains of the metal down the (respective) concentration gradients according to the usual diffusion
equation:

$$\frac{\partial c}{\partial t} = k \left\{ \frac{\partial^2 c}{\partial x^2} + \frac{\partial^2 c}{\partial y^2} + \frac{\partial^2 c}{\partial z^2} \right\},$$ (2-1)

where k is the (assumed constant) diffusivity of the species
with concentration $c(x,y,z,t)$. We use henceforth the subscript A for carbon and B for chrome, noting that both k_A
and k_B are strongly temperature dependent.

Diffusion of each species along the grain boundaries to
the reaction sites is relatively rapid (for evidence of this see
Perkins et al. [1973]), so that we may take the reaction, and
the ensuing thermodynamic equilibrium, to hold all along the
grain boundaries. We take this equilibrium to be of the form
(arguing that here activity is reasonably measured by concentration)

$$C_A C_B^\mu = K,$$ (2-2)

where μ is about three or four, and K is strongly temperature
dependent.

Since we combine the two species in a constant way, say
ν of C_B for 1 of C_A, we must have the fluxes of the species
at the grain boundaries in the same proportion, so that

$$k_A \frac{\partial C_A}{\partial n} = \nu k_B \frac{\partial C_B}{\partial n}.$$ (2-3)

Here ν is about three, and $\partial/\partial n$ means the derivative along an
inward pointing normal to the grain boundary. We now have two
diffusion equations (2-1) coupled by the two boundary conditions
(2-2) and (2-3). If we prescribe initial conditions in the grain
for the carbon and chrome concentrations then we have a well
posed mathematical problem (that is, a problem with just one
solution continuously depending on the various parameters). However $k_A/k_B \geq 10^6$! (In the language of ordinary differential

equations the problem is stiff.)

At ordinary temperatures the values of k_A, k_B are so small that nothing happens in our time-scales of interest. At annealing temperatures the steel is saturated with the initial concentrations of say

$$C_A \equiv 0.1 \quad \text{and} \quad C_B \equiv 15 \qquad (2-4)$$

throughout the grain at, say, time $t = 0$. At the working temperatures we have, for $t > 0$, $K < 0.1 \times 15^{\mu}$. It is this drop of K at $t = 0$ in (2-2) (which is the result of supersaturation of the initial concentrations at the lower temperature) which initiates the whole process.

3.   THE MATHEMATICAL MODEL AND THE SIMPLIFYING ANALYSIS

We consider the coupled diffusion of the species A and B in the metal grain D. The boundary of the grain is denoted by ∂D, and the x axis is chosen to coincide with the shortest diameter through the grain. We choose a length-scale so that the grain lies in $-1 \leqslant x \leqslant 1$. We choose a time-scale appropriate to the fast diffusing species A so that

$$\frac{\partial C_A}{\partial t} = \frac{\partial^2 C_A}{\partial x^2} + \frac{\partial^2 C_A}{\partial y^2} + \frac{\partial^2 C_A}{\partial z^2} \quad \text{in } D. \qquad (3-1)$$

Here $C_A(x,y,z,t)$ is the scaled concentration of species A so that

$$C_A(x,y,z,0) \equiv 1 \quad \text{in } D. \qquad (3-2)$$

We denote the diffusivity of species B on these length and time-scales by ε^2 (where $\varepsilon^2 \ll 1$). We scale its concentration $C_B(x,y,z,t)$ so that

$$\frac{\partial C_B}{\partial t} = \varepsilon^2 \left\{ \frac{\partial^2 C_B}{\partial x^2} + \frac{\partial^2 C_B}{\partial y^2} + \frac{\partial^2 C_B}{\partial z^2} \right\} \quad \text{in } D, \qquad (3-3)$$

and

$$C_B(x,y,z,0) \equiv 1 \quad \text{in } D. \qquad (3-4)$$

With these scalings we find that the equilibrium boundary condition (2-2) takes the form, for $\mu \simeq 4$,

$$C_A \, C_B^\mu = K < 1 \quad \text{on} \quad \partial D. \tag{3-5}$$

The flux boundary condition (2-3) becomes, for $\nu \simeq 3$,

$$\frac{\partial C_A}{\partial n} = \nu \varepsilon^2 \, \eta^{-1} \, \frac{\partial C_B}{\partial n} \quad \text{on} \quad \partial D, \tag{3-6}$$

where $\partial / \partial n$ is the inward normal derivative on the boundary and η is the ratio of the concentration scalings, with $\eta \simeq 0.01$ typically. Here K is set for $t > 0$, and it is inconsistent with (3-2,3-4). This sets off the whole process as C_A and/or C_B must decrease for $t > 0$.

Typically, for 17% stainless steels with 0.1% by weight of carbon worked at around $650°C$, we find that $10^{-10} < \varepsilon^2 < 10^{-6}$ and $0.1 < K < 0.9$. We note that

$$0 < \varepsilon \ll \varepsilon/\eta \ll 1 \tag{3-7}$$

It is ε^{-2} which is the very long timescale of the slowly diffusing species B, while it is $(\eta/\varepsilon)^2$ which yields the intermediate time-scale of practical interest.

The implications of these parameter values are as follows. Since $\varepsilon^2/\eta \ll 1$, we see that (3-6) implies that $\partial C_A/\partial n \simeq 0$ unless a steep boundary layer forms in the B concentration profile near ∂D. Thus for very small time it is B which must drop in concentration in order to satisfy (3-5). This produces a boundary layer of width approximately ε near ∂D for C_B when $t \simeq 1$.

Since $\partial C_B/\partial n$ is then large, approximately ε^{-1}, we have from (3-6) that $\partial C_A/\partial n \simeq \varepsilon/\eta \ll 1$ near ∂D. Thus there results a small, but continuing, drain of carbon across the boundary, and this continues up to a time $(\eta/\varepsilon)^2$ when the carbon concentration begins to fall across the whole grain.

We now quantify the preceding discussion. Our first task is to relate the flux of C_B on ∂D to the actual value of C_B on ∂D, and we achieve this in (3-11). Since it will turn out that the intermediate time-scale τ only is of interest, we let

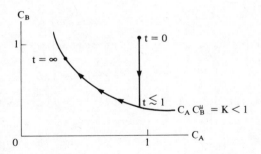

Figure 1. The locus of C_A, C_B as a function of time.

the concentration of species B on ∂D be defined by $B(\tau)$. The boundary layer approximation means that (3-3) is virtually

$$\frac{\partial C_B}{\partial t} \equiv \frac{1}{\beta} \frac{\partial C_B}{\partial \tau} = \varepsilon^2 \frac{\partial^2 C_B}{\partial n^2} , \qquad (3-8)$$

where $\partial/\partial n$ is rate of change along the inward normal and

$$t = \beta\tau , \quad \beta \gg 1 \gg \varepsilon . \qquad (3-9)$$

If n is distance along the normal inwards from ∂D, then

$$C_B = B(\tau) \text{ for } n = 0 \text{ and } C_B \to 1 \text{ as } n/\varepsilon\beta^{\frac{1}{2}} \to \infty . \quad (3-10)$$

As $\tau \to 0$ we have $C_B \to 1$ at all points inside D.

So we solve (3-8, 3-10) by taking Laplace Transforms of $1 - C_B$ in the time variable τ. Let $g(s, n)$ be the transform of $1 - C_B$ for $n > 0$, and $G(s)$ be the transform of $1 - B(\tau)$. Then (3-8) and (3-10) imply that

$$s g(s, n) = \beta\varepsilon^2 \frac{d^2 g}{dn^2} ; \quad \text{hence } g(s, n) = G(s) e^{-\sqrt{(s/\beta)}\,(n/\varepsilon)} .$$

But now we also know that

$$\text{Laplace Transform of } \frac{\partial C_B}{\partial n} \text{ on } \partial D = -\frac{dg}{dn} \text{ at } n = 0 ,$$

$$= \sqrt{\frac{s}{\beta}} \frac{1}{\varepsilon} G(s) \equiv s \frac{1}{\sqrt{s}} G(s) \frac{1}{\varepsilon\sqrt{\beta}} .$$

We invert this latest expression by using the Convolution Theorem, and remembering that s times the transform is equivalent to the τ derivative of the original. Thus we arrive at

$$\frac{\partial C_B}{\partial n}\bigg|_{n=0} = \frac{1}{\varepsilon\sqrt{\beta}} \frac{d}{d\tau} \frac{1}{\sqrt{\pi}} \int_0^\tau \frac{1-B(\hat{\tau})}{\sqrt{\tau-\hat{\tau}}}\, d\hat{\tau}\; ; \qquad (3\text{-}11)$$

we have expressed the flux in terms of the concentration of B on ∂D.

Our next task is to relate, on the τ timescale, the flux of A on ∂D to $A(\tau)$, the concentration of A on ∂D. We claim that, on the τ timescale, $\partial C_A/\partial n$ is fairly small and C_A is fairly constant over D. Thus we write

$$C_A = A(\tau) + C_1(x,y,z,t) \text{ in } D, \qquad (3\text{-}12)$$

where C_1 is smaller than $A(\tau)$.

Since $t = \beta\tau$, substitution of (3-12) in (3-1) gives

$$\frac{1}{\beta} \frac{dA}{d\tau} = \frac{\partial^2 C_1}{\partial x^2} + \frac{\partial^2 C_1}{\partial y^2} + \frac{\partial^2 C_1}{\partial z^2} - \frac{1}{\beta} \frac{\partial C_1}{\partial \tau} \text{ in } D. \qquad (3\text{-}13)$$

We claim that C_1 is of the size β^{-1}, and that $\partial C_1/\partial \tau$ is small. We wish to integrate (3-13) over D (using the Divergence Theorem), and we get, on ignoring the smaller $\beta^{-1}\,\partial C_1/\partial\tau$ term,

$$-\frac{1}{\beta} \frac{dA}{d\tau} \iiint_D dV = \iint_{\partial D} \frac{\partial C_1}{\partial n}\, ds = \iint_{\partial D} \frac{\partial C_A}{\partial n}\, ds. \qquad (3\text{-}14)$$

Note that $A(\tau)$ is independent of position (so that $\partial A(\tau)/\partial n = 0$ and $\partial C_A/\partial n = \partial C_1/\partial n$).

We let $|D| = $ volume of D and $|\partial D| = $ surface area of ∂D. Then, since (3-11) forces $\partial C_B/\partial n$ to be independent of position on ∂D, (3-6) forces $\partial C_A/\partial n$ to be independent of position on ∂D. So (3-14) simplifies to

$$\frac{dA}{d\tau} = -\beta \frac{|\partial D|}{|D|} \frac{\partial C_A}{\partial n} \text{ on } \partial D; \qquad (3\text{-}15)$$

we have related the flux and the concentration of A on ∂D.

Finally we use (3-6) to relate in (3-11),(3-15) the fluxes of A and B on ∂D, so that

$$\frac{dA}{d\tau} = -\beta \frac{|\partial D|}{|D|} \frac{\nu\varepsilon^2}{\eta} \frac{1}{\varepsilon\sqrt{\beta}} \frac{d}{d\tau} \frac{1}{\sqrt{\pi}} \int_0^\tau \frac{1-B(\hat{\tau})}{\sqrt{\tau-\hat{\tau}}}\, d\hat{\tau}. \qquad (3\text{-}16)$$

Now (3-5), which amounts to $A(\tau) B(\tau)^{\mu} = K$, may be used to elimi-nate $B(\tau)$ in (3-16), so that integration of (3-16) with respect to τ from $\tau = 0$, with $A(0) = 1$, shows that

$$1 - A(\tau) = \int_0^{\tau} \frac{1 - [K/A(\hat{\tau})]^{1/\mu}}{\sqrt{\tau - \hat{\tau}}} d\tau \sqrt{\frac{\beta}{\pi}} \frac{\nu \varepsilon}{\eta} \frac{|\partial D|}{|D|} \ . \quad (3-17)$$

4.   THE INTEGRAL EQUATION AND ITS NUMERICAL SOLUTION

We are free to choose β in (3-17), and this choice deter-mines the intermediate time-scale $\tau = \beta^{-1} t$. So to reduce our integral equation to its simplest form we choose

$$\beta = \pi \left[\frac{\eta |D|}{\nu \varepsilon |\partial D|} \right]^2 \gg 1, \quad (4-1)$$

and then our integral equation is

$$1 - A(\tau) = \int_0^{\tau} \frac{1 - [K/A(\hat{\tau})]^p}{\sqrt{\tau - \hat{\tau}}} d\hat{\tau} , \quad (4-2)$$

where $p = \mu^{-1} \simeq 0.3$. We have the one parameter K with $0 < K < 1$. Then $A(\tau)$ starts at 1 and decreases monotonically to K as τ increases to infinity. We wish to find numerical approximations to $A(\tau)$ over the whole range of τ.

Clearly the nonlinearity in the integral is a difficulty. A simple explicit numerical method to avoid this proceeds as follows. Define $A_0 = 1$, $\tau_0 = 0$, $\tau_i = i \Delta \tau$ and A_i for $i = 1, 2, 3, \ldots$ by

$$1 - A_i = \sum_{j=1}^{i} 2 \left\{ 1 - [K/A_{j-1}]^p \right\} \left\{ \sqrt{\tau_i - \tau_{j-1}} - \sqrt{\tau_i - \tau_j} \right\} . \quad (4-3)$$

Here A_i approximates $A(\tau_i)$ for $i \geqslant 1$. The integral has been evaluated by the product Euler rule using only previously calcu-lated approximations A_j, $j < i$, for $A(\tau)$. However this simple scheme is not very accurate. Asymptotically the errors can be of the size $O(\Delta \tau)^{\frac{1}{2}}$.

In the remainder of this section we show that $A'(\tau)$ has $1/\sqrt{\tau}$ behaviour near $\tau = 0$ and $A(\tau)$ has $1/\sqrt{\tau}$ decay to K as

$\tau \to 0$; and then we introduce transformations of the independent variable of integration to yield our basic result.

An effective way to approximate (4-2) is to introduce a graded mesh and then to approximate the integral by a simple trapezium rule of integration. This procedure was found to give about 10^{-3} relative accuracy over the whole range $0 < \tau < \infty$ when 100 points τ_i were unevenly placed along the τ axis by means of the rule $\tau_i = \tan^2\theta_i$ where $\theta_i = i\pi/2(N+1)$ for $1 \leqslant i \leqslant N$. We divide the region of integration $0 \leqslant \hat{\tau} \leqslant \tau_i$ by the knots τ_j for $j = 0, \ldots, i$ and use a piecewise linear approximation to $1 - K^p A^{-p}(\hat{\tau})$, evaluating the resulting integral exactly. The casual reader can now go directly to the discussion of results in Section 5. The remainder of this section describes why the above method works well.

Our major problem in getting an efficient numerical solution is the weakly singular integrand. In order to understand the effect of the term $1/\sqrt{\tau - \hat{\tau}}$, we look at the behaviour of the solution $A(\tau)$ for $\tau \to 0$ and for $\tau \to \infty$. When $\tau \to 0$ (4-2) becomes approximately

$$1 - A(\tau) \simeq \int_0^\tau \frac{1 - K^p}{\sqrt{\tau - \hat{\tau}}} \, d\hat{\tau} = (1 - K^p) 2\tau^{\frac{1}{2}} \; ;$$

that is,

$$A(\tau) \simeq 1 - 2(1 - K^p) \tau^{\frac{1}{2}} \quad \text{as} \quad \tau \to 0 . \qquad (4\text{-}4)$$

For $\tau \to \infty$ we must have $A(\tau) \to K$ so that the integral remains finite. We write $A(\tau) = K + a(\tau)$, and then (4-2) becomes approximately

$$1 - K \simeq \int_0^\tau \frac{1 - \left[\dfrac{K}{K + a(\hat{\tau})}\right]^p}{\sqrt{\tau - \hat{\tau}}} \, d\hat{\tau} \quad \text{for} \quad \tau \gg 1,$$

$$\simeq \int_0^\tau \frac{p}{K} \frac{a(\hat{\tau})}{\sqrt{\tau - \hat{\tau}}} \, d\hat{\tau} \quad \text{for} \quad \tau \gg 1 .$$

Using Laplace Transforms and the Convolution Theorem it follows that $a(\tau) = a\tau^{-\frac{1}{2}}$, and we find that

$$A(\tau) \simeq K\left\{1 + \frac{1-K}{\pi p}\tau^{-\frac{1}{2}}\right\} \quad \text{for} \quad \tau \gg 1. \tag{4-5}$$

Brunner [1983] describes methods for coping with the singularity as $\tau \to 0$. He uses nonpolynomial spline collocation, which is complicated. A simpler and numerically more efficient method is to transform variables so that we can use standard low order polynomial spline collocation (piecewise constant or linear).

The appearance of $\sqrt{\tau}$ in both (4-4),(4-5) suggests the use of the transformation $\tau = s^2$ (see Noble [1964], who used this for $\tau \to 0$). The transformation $\hat{\tau} = \tau \sin^2\phi$ removes the singularity in the denominator of the integral, so that (4-2) becomes

$$1 - A(s^2) = \int_0^{\pi/2} \left\{\frac{1 - [K/A(s^2\sin^2\phi)]^p}{s\cos\phi}\right\} s^2\, 2\sin\phi\,\cos\phi\,d\phi. \tag{4-6}$$

It is more convenient to write $s = \tan\theta$ so that $0 < s < \infty$ becomes $0 < \theta < \frac{\pi}{2}$. Setting $A(s^2) \equiv A(\tan^2\theta) = u(\theta)$ and defining ϕ^* by $\tan\phi^* = \tan\theta\,\sin\phi$ equation (4-6) becomes

$$1 - u(\theta) = \int_0^{\pi/2} \tan\theta\left\{1 - K^p A^{-p}(\tan^2\theta\,\sin^2\phi)\right\} 2\sin\phi\,d\phi$$

$$= 2\tan\theta \int_0^{\pi/2}\left\{1 - K^p u^{-p}(\phi^*)\right\}\sin\phi\,d\phi$$

$$\equiv \int_0^{\pi/2} M(\theta,\phi,K)\,d\phi. \tag{4-7}$$

The transformed integral equation (4-7) has a regular solution $u(\theta)$ which decreases monotonically from $u(0) = 1$ to $u(\frac{\pi}{2}) = K$ as θ increases from 0 to $\frac{\pi}{2}$. The only possibility of bad behaviour arises as $\theta \to \frac{\pi}{2}$ but as $\theta \to \frac{\pi}{2}$, $\phi^* \to \frac{\pi}{2}$ ($\phi > 0$) and $M(\frac{\pi}{2},\phi,K) = 2(1-K)\pi^{-1}$. Note that $M(\theta,0,K) = 0$ so that M is discontinuous at $\theta = \frac{\pi}{2}$, $\phi = 0$.

It is now possible to improve upon the simple explicit scheme (4-3).

Let $0 = \theta_0 < \theta_1 < \ldots < \theta_{N+1} = \frac{\pi}{2}$. Then setting $\theta = \theta_i$, $0 \leqslant i \leqslant n$, in (4-7) we have

$$1 - u(\theta_i) = \int_0^{\pi/2} \tan\theta_i \left\{ 1 - K^p u^{-p} \left(\tan^{-1}(\tan\theta_i \sin\phi) \right) \right\} 2\sin\phi \, d\phi.$$

$$(4\text{-}8)$$

Now partition the interval $0 \leqslant \phi \leqslant \frac{\pi}{2}$ by the points ϕ_{ij} defined by

$$\sin\theta_{ij} = \frac{\tan\theta_j}{\tan\theta_i}, \qquad 0 \leqslant j \leqslant i. \qquad (4\text{-}9)$$

Then (4-8) becomes

$$1 - u(\theta_i) = \int_0^{\phi_{ii}} \tan\theta_i \left\{ 1 - K^p u^{-p} \left(\tan^{-1}(\tan\theta_i \sin\phi) \right) \right\} 2\sin\phi \, d\phi.$$

Approximating the integral by numerical quadrature gives

$$1 - u_i = \sum_{j=0}^{i} a_{ij}(u_j) \qquad (4\text{-}10)$$

where u_i denotes an approximation to $u(\theta_i)$ and the a_{ij} depend upon the integration formula employed.

To determine u_i it is necessary to solve an equation of the form

$$u_i - a_{ii}(u_i) = g \qquad (4\text{-}11)$$

where g contains previously computed values u_j, $0 \leqslant j \leqslant i-1$. In general, (4-11) is nonlinear and an iterative procedure is required to determine u_i.

Now $u(\tan^{-1}(\tan\theta_i \sin\phi_{ij})) \equiv u(\theta_j)$, therefore, approximating by u_j for $\phi_{ij} \leqslant \phi^* \leqslant \phi_{ij+1}$ we have in (4-10)

$$a_{ij}(u_j) = \tan\theta_i \left\{ 1 - K^p u_j^{-p} \right\} 2 \left\{ \cos\phi_{ij} - \cos\phi_{ij+1} \right\},$$

$$0 \leqslant j \leqslant i-1, \quad 1 \leqslant i \leqslant N$$

$$a_{ii}(u_i) = 0, \qquad\qquad\qquad\qquad 1 \leqslant i \leqslant N.$$

This gives the simplest (explicit) scheme. If instead we approximate $u(\phi^*)$ over $[\phi_{ij}, \phi_{ij+1}]$ by the Lagrange polynomial of degree 1 given by

$$u_j \left\{ \frac{\phi_{ij+1} - \phi^*}{\phi_{ij+1} - \phi_{ij}} \right\} + u_{j+1} \left\{ \frac{\phi^* - \phi_{ij}}{\phi_{ij+1} - \phi_{ij}} \right\} \qquad (4\text{-}12)$$

then (4-10) yields the following implicit scheme for $1 \leqslant i \leqslant N$,

$$\frac{1 - u_i}{2 \tan \theta_i} = 1 + K^p \sum_{j=0}^{i-1} \left\{ u_{j+1}^{-p} \left(\cos \phi_{ij+1} - S_{ij} \right) - u_j^{-p} \left(\cos \phi_{ij} - S_{ij} \right) \right\}$$

$$(4\text{-}13)$$

where

$$S_{ij} = \frac{\sin \phi_{ij+1} - \sin \phi_{ij}}{\phi_{ij+1} - \phi_{ij}} \quad .$$

This defines $a_{ij}(u_j)$ for $0 \leqslant j \leqslant i$, so that with $\phi_{i0} = 0$, $\phi_{ii} = \pi/2$ and $\Delta \phi = \pi/2 - \phi_{ii-1}$, we have

$$a_{ii}(u_i) = 2 \tan \theta_i \, K^p u_i^{-p} \left\{ \frac{\cos \Delta \phi - 1}{\Delta \phi} \right\} . \qquad (4\text{-}14)$$

Thus we have, for each $i \geqslant 1$,

$$u_i + 2 \tan \theta_i \, K^p u_i^{-p} \left\{ \frac{\cos \Delta \phi - 1}{\Delta \phi} \right\} = \text{known} . \qquad (4\text{-}15)$$

Finally, in order to solve (4-15) it was found in practice that only a single Newton step sufficed when the initial guess for u_i^{-p} was taken to be u_{i-1}^{-p}. For $i \geqslant 1$ (4-15) becomes

$$u_i + 2 \tan \theta_i \, p \, K^p u_{i-1}^{-p-1} \left\{ \frac{1 - \cos \Delta \phi}{\Delta \phi} \right\} u_i = \text{known} . \qquad (4\text{-}16)$$

This latter scheme (4-16) can be solved directly for u_i, and the solution process displays no sign of numerical instability as $\theta_i \to \pi/2$ (i increases to $N+1$; $\tau \to \infty$). (In contrast, the method based on (4-11) would appear to be numerically unstable as $\theta_i \to \pi/2$.) In practice, one hundred evenly spaced knots θ_i in the interval $[0, \pi/2]$, together with the formula (4-16), are sufficient to give about three decimal place accuracy (providing K is not too small) of the solution u_i of (4-8). One hundred evaluations of (4-16) uses very little computer power. The

following table gives $u(\theta_i)$ for $\theta_i = 0.706858$ for $K = 0.1, 0.5,$
0.9. Note that $u \equiv 1$ for $K = 1$, that $u(0) \equiv 1$, and that $u(\pi/2) \equiv K$.
See Norbury & Stuart [1985] for a convergence proof for the above
numerical method — the convergence ratios indicate 3/2 approximately.

Table 1 Numerical values for $u(0.706858)$

N \ K	0.1	0.5	0.9
80	0.371314	0.770255	0.962498
160	0.372020	0.770476	0.962532
320	0.372290	0.770558	0.962545
640	0.372390	0.770588	0.962550

5.   DISCUSSION AND RESULTS

In predicting the ageing behaviour of stainless steel
used in the operating loops of sodium cooled nuclear reactors we
have gone through several stages. First we modelled (in Section
2) the physical processes by a pair of nonlinearly coupled diffusion equations. Then we nondimensionalized and rescaled these
equations in Section 3, and discovered the presence of the two
small parameters ε and η. We exploited the smallness of these
parameters in a simplifying mathematical analysis that led to a
nonlinear singular Volterra integral equation.

In Section 4 we examined the behaviour of the solution of
the integral equation, and this suggested a series of transformations that led to a more regular, but nonstandard, integral
equation. Finally we suggested some numerical methods for the
evaluation of the solution of this integral equation. The solution $u(\theta)$ of the transformed integral equation is shown, for
various K, in Figure 2. Note that u decreases monotonically
as θ increases and as K decreases. The nonlinearity of the
integrand ensures that $u(\theta) \to K$ as $\theta \to \pi/2$.

In Figure 3 we show the values of $A(\tau)$ and of $B(\tau) =$
$[K/A(\tau)]^p$ for various values of K. The value of $A(\tau)$ gives

J. NORBURY

the carbon concentration at the grain boundary on the intermediate time-scale $t = \beta\tau$ ($\beta \simeq 300$ is typical). The variation of
$C_A - A(\tau)$ throughout the grain is typically less than 1%. The
value of $B(\tau)$ gives the surface concentration of chrome on the
intermediate time-scale. In the middle of the grain C_B remains
at its initial level during this time period.

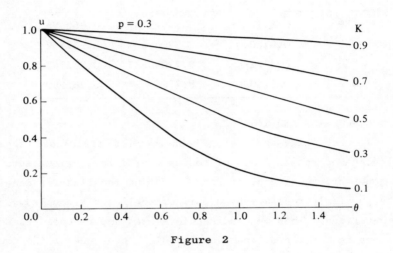

Figure 2

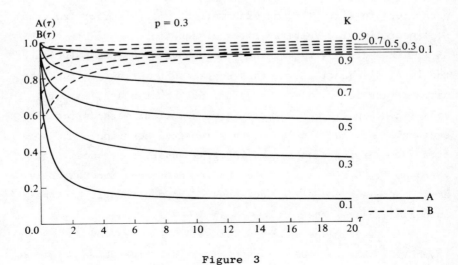

Figure 3

The chrome concentration C_B changes in a boundary layer extending a scale distance η inwards from the grain boundary. An idea of the variation of C_B along an inward normal to the boundary may be obtained from the formula

$$C_B = B(\tau) + [1 - B(\tau)] \ \text{erf} \ (n / \varepsilon \sqrt{\beta t}).$$

Thus, for all times, the profiles of the carbon and chrome concentrations are known throughout D to an accuracy of better than 1%.

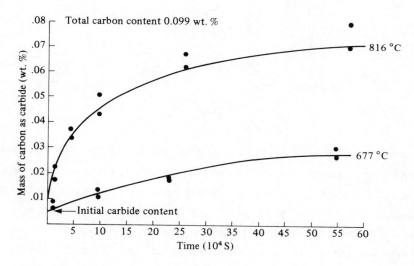

Figure 4

We are now in a position to compare the predictions of this model with experimental results. Figure 4 shows the mass of carbon precipitated after various times from the experiments of Lai & Meshkat [1977] and Bendure *et al*. [1961] compared with suitably scaled curves of our $1 - A(\tau)$. Note that the intermediate time-scale is predicted, as are the qualitative behaviours (even the quantitative agreement is reasonable).

Finally in Figure 5 we compare our predictions of the chrome boundary layer with some experimental results of Pande *et al*. [1977]. The qualitative agreement of our model with all

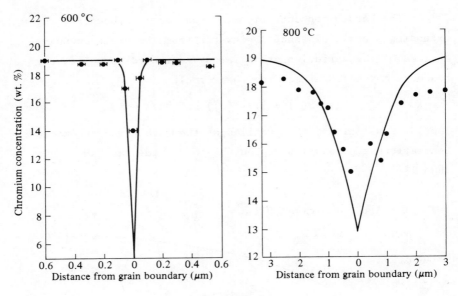

Figure 5

of these experimental results is good. Considering the enormous simplifications of the model the quantitative agreement is surprising. In fact, by altering our assumptions on grain dimensions and activity constants we could improve this quantitative agreement. However we think it is sufficient to say that the agreement with experimental results justifies the original model and its basic predictions.

REFERENCES

Bendure, R.J., Ikenberry, L.C. and Waxweiler, J.H. 1961, Quantity and form of carbides in austenitic and precipitation hardening stainless steels, *Trans. Met. Soc. AIME*, **221**, 1032-1062.

Brunner, H. 1983, Nonpolynomial spline collocation for Volterra equations with weakly singular kernels, *SIAM J. Numer. Anal.* **20**, (6), 1106-1119.

Coates, D.E. 1972, Diffusion controlled precipitate growth in ternary systems I, *Met. Trans.* **3**, 1203-12.

Coates, D.E., 1973, Diffusion controlled precipitate growth in ternary systems II, *Met. Trans.* **4**, 1077-86.

Goldstein, J.I. and Randich, E. 1977, Variation of interface compositions during diffusion controlled precipitate growth in ternary systems, *Met. Trans.* **8A**, 105-9.

Kircaldy, J.S. 1958, Precipitate growth in an infinite matrix in ternary systems, *Can. J. Phys.* **36**, 907-16.

Lai, J.K. and Meshkat, M. 1977, Carbide precipitation in stainless steel, CEGB Report RD/L/N22/77, 1-33.

Markham, G. 1980, The modelling of carbide formation in stainless steel in fast reactors, CEGB Report RD/L/N151/80, 1-13.

Noble, B. 1964, The numerical solution of nonlinear integral equations, *Nonlinear Integral Equations*. (P.M. Anselone, Ed.) 215-318.

Norbury, J. and Stuart, A.M. 1985, Singular nonlinear Volterra integral equations, (submitted to *SIAM J. Numer. Anal.*)

Pande, C.S., Suenage, M., Vyas, B., Isaacs, H.S. and Harling, D.F. 1977, Direct evidence of chromium depletion near the grain boundaries in sensitized stainless steels, *Scripta Metallurgica* **11**, 681-684.

Perkins, R.A., Padgett, R.A. and Tunali, N.K. 1973, Tracer diffusion of iron and chrome in austenitic alloy — stainless steel, *Met. Trans.* **4**, 2535-2540.

Skyrme, G. and Norbury, J. 1980, Kinetic of $M_{23}C_6$ carbide growth in type 316 stainless steel, CEGB Report RD/B/N4943/80, 1-42.

Snyder, R.B., Natesan, K. and Kassner, T.F. 1973, Precipitate growth in ternary systems, *J. Nucl. Mat.* **50**, 259.

Strawström, C. and Hillert, M. 1969, Chrome diffusion in stainless steel, *J. Iron and Steel Inst.*, **207**, 77-85.

Tedmon, C.S., Vermilyea, D.A. and Rosolowski, J.H. 1971, Diffusion controlled precipitate growth in ternary systems, *J. Electrochem. Soc.* **118**, 192-201.

9

The development of mathematical models in welding and their numerical solution

J. G. ANDREWS AND R. E. CRAINE

1. INTRODUCTION

There is an increasing tendency towards the use of mathematical models of welding processes as a tool towards improving weld quality, especially for automatic welding procedures. A wide variety of welding techniques are used today, the choice being dictated by the particular application. We concentrate here on a fusion welding process known as TIG (Tungsten electrode with an Inert Gas shield), which is employed extensively in the manufacture of boiler tubes and other heavy engineering plant. In the TIG process a high current electric arc is moved slowly along the line of contact between the two components to be joined. Melting occurs in the vicinity of the arc and the weld is formed when the molten weld pool freezes after the arc has moved away (see Figure 1).

The first models to be developed concentrated on the heat conduction processes in the components being welded and such solutions for both steady and unsteady state conditions can be found in Carslaw & Jaeger [1959]. These solutions have proved to be very popular with metallurgists, since they are simple and possess certain features which are in agreement with observation, but they suffer from a major limitation — the neglect of fluid motion in the pool. Experimental evidence for vigorous fluid

motion in the weld pool is overwhelming and account must be
taken of this motion if accurate predictions of weld pool shape
are to be made.

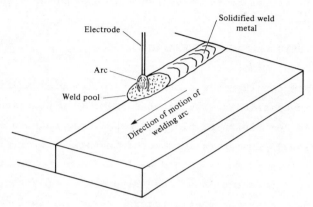

Figure 1
Arc welding of two plates with a moving heat source

It is widely thought that electromagnetic body forces
inside the pool are the principal cause of the flow motion;
their importance being first demonstrated by Woods & Milner
[1966] in a series of experiments using liquid mercury. Theoretical attempts to describe the fluid motion caused by these
electromagnetic forces have ignored thermal effects and have
assumed that the fluid is either semi-infinite or inside a con-
tainer of known shape. The developers of welding models have
been split therefore into two separate groups: the heat
conduction school which ignores motion in the pool and the fluid
motion researchers who omit heat flow. Clearly to predict
accurately the position of the weld pool boundary it is necessary to consider both aspects. In this Chapter we attempt to
do this by solving the fluid and heat flow equations simultaneously. The method is inevitably numerical, the key to it
being the use of body-fitted co-ordinates to transform the shape
of the boundary obtained at each iteration onto a hemisphere,
within which a numerical method used by Atthey [1980] is adapted

to solve for the fluid flow. The convective heat flow equation is then solved in this transformed region and the position of the melting isotherm is recalculated. The iterations are continued until satisfactory convergence of the boundary shape is achieved.

2.   THE PHYSICAL AND MATHEMATICAL MODEL

2.1   *Physical Model*

We now consider the particular problem of a stationary axisymmetric distributed source of current (and heat) on the surface of a semi-infinite block of metal (the magnitudes of the non-dimensional parameters introduced in this Chapter will refer to steel). This produces a weld pool of finite extent, the shape of which is to be established (see Figure 2). Only the expectedly dominant electromagnetic forcing terms are included in our model but their appearance might suggest that the full magneto-hydrodynamic equations must be solved. Fortunately, in real welding situations the magnetic Reynolds' number is $O(10^{-3})$ and the back e.m.f. effects can be neglected. This is a crucial simplification since the electromagnetic field quantities are then independent of the fluid flow and, in the steady state, an electric potential exists which satisfies Laplace's equation.

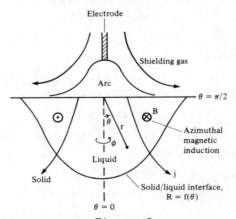

Figure 2

Schematic section of axisymmetric weld pool produced by the TIG (Tungsten-inert-gas) welding process.

A (fluid) Reynold's number of $O(10^2)$, typically, indicates that the velocity gradients are likely to be fairly large near the solid/liquid interface. However, since the ratio of the characteristic thicknesses of the viscous and thermal boundary layers for external flows is $O(Pr^{\frac{1}{2}})$ (Curle [1962]) and the Prandtl number, Pr, is $O(10^{-1})$ in welding situations we do not expect the temperature gradient at the boundary of our internal flow to depend critically on the details of the velocity field in this region. The shape will be sensitive though to the bulk motion of the fluid in the body of the pool.

2.2   *Mathematical Model*

Using the stream function-vorticity formulation the steady state equations of motion for the fluid are, in conservative form and expressed in nondimensional variables,

$$\tfrac{1}{2} R^2 \sin^2 \theta \nabla \cdot \left(\frac{\boldsymbol{V}\Omega}{R^2 \sin^2 \theta} \right) = 2KF + \emptyset \Omega \qquad (2\text{-}1)$$

and

$$\emptyset \Psi = -\Omega \,, \qquad (2\text{-}2)$$

where ∇ is referred to the nondimensional spherical polar coordinates (R, θ, ϕ) and the Stokes operator $\emptyset$ is defined by

$$\emptyset \equiv \frac{\partial^2}{\partial R^2} + \frac{1}{R^2} \left(\frac{\partial^2}{\partial \theta^2} - \cot \theta \, \frac{\partial}{\partial \theta} \right) \,.$$

An amplified version of this section can be found in Craine & Andrews [1984]. In equations (2-1) and (2-2) Ω and Ψ represent the non-zero vorticity component and stream function respectively, $\boldsymbol{V}$ denotes the fluid velocity and K is the non-dimensional constant introduced by Sozou [1971]. For a Gaussian input of current I on the surface $\theta = \tfrac{1}{2}\pi$ the appropriate form of F, the only non-zero component of the curl of the electromagnetic stirring, can be shown to be

$$F = \int_0^\infty \int_0^\infty \exp \left\{ -\tfrac{1}{4}(\alpha^2 + \alpha'^2) R_0^2 - (\alpha + \alpha') R \cos \theta \right\}$$

$$\alpha J_1(\alpha R \sin \theta) \, J_1(\alpha' R \sin \theta) \, d\alpha \, d\alpha' \qquad (2\text{-}3)$$

where $J_1(\cdot)$ is the first order Bessel function of the first kind and R_0 denotes the nondimensional decay radius of the Gaussian current source (so that on the top surface at $R = R_0$ the input current is $\frac{1}{e}$ times its maximum value).

The equation of steady state convective heat flow may be written

$$\nabla^2 \Phi = \tfrac{1}{2} Pr \, (\boldsymbol{V} \cdot \nabla) \, \Phi, \qquad (2-4)$$

where Φ is the nondimensional quantity defined from the temperature T by $\Phi = (T - T_{(a)})/(T_{(m)} - T_{(a)})$; $T_{(a)}$ and $T_{(m)}$ denoting the ambient and melting temperatures respectively. Clearly $V = 0$ in the solid in which case (2-4) then reduces to Laplaces' equation.

2.3   Body Fitted Coordinates

Equations (2-1), (2-2) and (2-4) form a set of coupled nonlinear equations with a free interface. We shall follow Thompson *et al*. [1974] and introduce body fitted coordinates, a new coordinate system in which the solid/liquid interface becomes a constant coordinate surface thereby enabling us to use finite difference methods without having to introduce inaccuracies through interpolative schemes near the interface. In our problem we transform (R, θ) space into (ξ, θ) space through the simple relation

$$\xi(R, \theta) = R/f(\theta) \qquad (2-5)$$

where $R = f(\theta)$ is the equation of the unknown solid/liquid interface. This transformation maps the pool onto the hemisphere of unit radius but complicates the partial differential equations for fluid and heat flow. In (ξ, θ) space equations (2-1) and (2-2) become

$$\frac{\xi^2 \sin^2 \theta}{2f} \, \tilde{\nabla} \cdot \left(\frac{\boldsymbol{W}\Omega}{\xi^2 f^2 \sin^2 \theta} \right) = 2K\tilde{F} + \tilde{\mathcal{D}}\,\Omega \qquad (2-6)$$

and

$$\tilde{\mathcal{D}} \, \Psi = -\Omega, \qquad (2-7)$$

in which $\tilde{\nabla}$ and $\tilde{F}$ denote ∇ and F in (ξ, θ) space and where, using subscripts ξ and θ to denote partial differentiation with respect to those variables

$$W = \left(\frac{\Psi_\theta}{\xi^2 \sin \theta}, \quad -\frac{\Psi_\xi}{\xi^2 \sin \theta}, \quad 0 \right),$$

$$\tilde{\Omega} \equiv \beta^{(1)} \partial_{\xi\xi} + \beta^{(2)} \partial_{\xi\theta} + \beta^{(3)} \partial_{\theta\theta} + \beta^{(4)} \partial_\xi + \beta^{(5)} \partial_\theta,$$

$$\beta^{(1)} = (f^2 + (f_\theta)^2)/f^4, \quad \beta^{(2)} = -2 f_\theta/(\xi f^3), \quad \beta^{(3)} = 1/(\xi f)^2,$$

$$\beta^{(4)} = \left\{ \frac{2(f_\theta)^2}{f} - f_{\theta\theta} + (\cot \theta) f_\theta \right\} \Big/ (\xi f^3), \quad \beta^{(5)} = -\cot \theta/(\xi f)^2.$$

Similarly, the equation for heat flow (2-4) becomes:

$$\beta^{(1)} \Phi_{\xi\xi} + \beta^{(2)} \Phi_{\xi\theta} + \beta^{(3)} \Phi_{\theta\theta} + \beta^{(6)} \Phi_\xi - \beta^{(5)} \Phi_\theta = \beta^{(7)} (\Psi_\theta \Phi_\xi - \Psi_\xi \Phi_\theta),$$

$$(2\text{-}8)$$

where

$$\beta^{(6)} = \left\{ \frac{2(f_\theta)^2}{f} - f_{\theta\theta} + 2f - (\cot \theta) f_\theta \right\} \Big/ (\xi f^3),$$

$$\beta^{(7)} = \frac{Pr}{2\xi^2 f^2 \sin \theta}.$$

Note that for clarity the dependence of f on θ is not shown.

2.4   *Boundary Conditions*

The boundary of the pool must everywhere coincide with a streamline which we take to be $\Psi = 0$. Moreover, the free surface is assumed flat and to have zero tangential shear and as a result we require

$$\Psi = \Omega = 0 \quad \text{on} \quad \theta = \tfrac{1}{2}\pi, \quad 0 \leqslant \xi \leqslant 1. \qquad (2\text{-}9)$$

(The assumption of a flat free surface seems reasonable in view of the detailed discussion in Atthey [1980].) On the axis of symmetry, which is also a streamline, we have

$$\Psi = \Omega = \Omega_\theta = 0 \quad \text{on} \quad \theta = 0, \quad 0 \leqslant \xi \leqslant 1, \qquad (2\text{-}10)$$

and since the origin is a stagnation point and the vorticity must be single-valued there we assume further that

$$\Psi = \Psi_\xi = \Omega = \Omega_\xi = 0 \quad \text{at} \quad \xi = 0, \quad 0 \leqslant \theta \leqslant \tfrac{1}{2}\pi \qquad (2\text{-}11)$$

Both velocity components being zero on the solid/liquid interface, $\xi = 1$, implies

$$\Psi = \Psi_\xi = \Omega + \beta^{(1)}\Psi_{\xi\xi} = 0 \quad \text{on} \quad \xi = 1, \quad 0 \leqslant \theta \leqslant \tfrac{1}{2}\pi. \quad (2\text{-}12)$$

For the boundary conditions on the temperature it is appropriate
to prescribe on the upper surface, $\theta = \tfrac{1}{2}\pi$, a Gaussian heat flux
input, similar to that of the current distribution (Nestor [1962]).
In (ξ, θ) space this input condition becomes

$$\frac{1}{\xi}\Phi_\theta = \frac{f_\theta}{f}\Phi_\xi + \frac{q}{\xi_0^2}f\exp\left\{-(\xi f/\xi_0)^2\right\} \quad \text{on} \quad \theta = \tfrac{1}{2}\pi, \quad 0 \leqslant \xi \leqslant \infty,$$
$$(2\text{-}13)$$

where $\xi_0 = R_0 f(\tfrac{1}{2}\pi)$ is the scaled decay radius and q is a non-
dimensional constant directly proportional to the total power
dissipated by the arc on the surface of the material. On the
axis of symmetry we have

$$\Phi_\theta = 0 \quad \text{on} \quad \theta = 0, \quad 0 \leqslant \xi \leqslant \infty. \quad (2\text{-}14)$$

In our numerical treatment we cannot apply the exact conditions
on Φ at infinity, so the latter is replaced in (R, θ) space by
a hemisphere of large radius, which transforms in (ξ, θ) space
to $\xi = \xi^{(h)}(\theta)$, say. The ranges in (2-13) and (2-14) then change
to $0 \leqslant \xi \leqslant \xi^{(h)}(\tfrac{1}{2}\pi)$ and $0 \leqslant \xi \leqslant \xi^{(h)}(0)$ respectively. We
assumed that on the outer hemisphere the temperature is that
given by a point source of heat (of the same strength) at the
origin, and this leads to

$$\Phi = \frac{q}{2\xi f} \quad \text{on} \quad \xi = \xi^{(h)}(\theta), \quad 0 < \theta < \tfrac{1}{2}\pi. \quad (2\text{-}15)$$

Various known exact heat conduction solutions for non-hemispheri-
cal bodies suggest that the error incurred by using (2-15) is less
than 5%. Such accuracy is sufficient for our purposes since we
are seeking large scale efforts.

Figure 3 describes the region in which a solution is to
be sought together with the appropriate boundary conditions.
Clearly the conditions on Ψ and Ω are only applied on ∂D_L
but those on Φ need applying on $\partial(D_L + D_S)$.

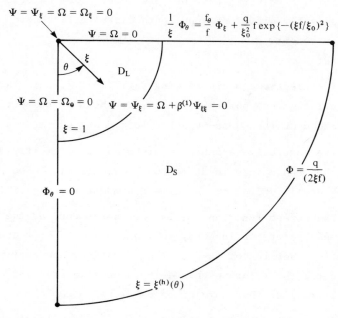

$$\Psi = \Psi_\xi = \Omega = \Omega_\xi = 0$$

$$\Psi = \Omega = 0$$

$$\frac{1}{\xi}\,\Phi_\theta = \frac{f_\theta}{f}\,\Phi_\xi + \frac{q}{\xi_0^2}\,f\exp\{-(\xi f/\xi_0)^2\}$$

$$\theta \qquad \xi \qquad D_L$$

$$\Psi = \Omega = \Omega_\theta = 0 \qquad \Psi = \Psi_\xi = \Omega + \beta^{(1)}\Psi_{\xi\xi} = 0$$

$$\xi = 1$$

$$D_S$$

$$\Phi = \frac{q}{(2\xi f)}$$

$$\Phi_\theta = 0$$

$$\xi = \xi^{(h)}(\theta)$$

Figure 3

**The regions in (ξ, θ) space in which solution is
required with appropriate boundary conditions.**

3.   NUMERICAL METHODS

To determine the fluid flow and temperature profiles we
solve the system derived in Section 2 using, in contrast to
the following chapter, a finite-difference procedure. Since a
full description of the numerical methods appears elsewhere
(Craine and Andrews [1984]) we shall provide here only a brief
outline of the approach.

Procedure

1.   Initial values for $f(\theta)$, Ψ, Ω and Φ are given. The
values used are either

(a) $f(\theta) = 1$, $\Psi = \Omega = 0$ and $\Phi = 2 - \xi$ in D_L with Φ
in D_S being calculated from the temperature distribution
due to a point source of heat at the origin.

or   (b)   a solution that has previously been calculated for
          slightly different input parameters (c.f. continuation
          approach).
          Set $n := 0$.

2.   The electromagnetic forcing term (2-3) is calculated at
          all the mesh points in D_L using the NAG subroutine
          based on the Gauss–Laguerre method of quadrature.

3.   The vorticity variable, Ω, is updated explicitly at half
          the mesh points in D_L through essentially a Du Fort-
          Frankel scheme for the time dependent version of (2-6).

4.   The stream function, Ψ, is then determined at the same
          mesh points as in 3 by solving, using SOR , a discre-
          tised version of (2-7). The iterations are continued
          until the relative change in successive iterates of Ψ
          is smaller than some given parameter (say ε).

5.   Steps 3 and 4 are repeated at the other half of the mesh
          points in D_L .

6.   Are the relative changes in the iterates of Ω and Ψ at
          all mesh points during one sweep through steps $3-5$ below
          some given tolerances (say ε_Ω and ε_Ψ respectively)? If
          so, then proceed to 7; if not, return to 3.

7.   Discretise (2-8) and solve for Φ by SOR , stopping the
          iterations only when the relative changes in Φ are less
          than ε_Φ, say.
          Set $n := n + 1$,

8.   The new position $R = f_n(\theta)$ of the solid/liquid interface
          is estimated from the most recent Φ iterates by fitting
          local parabolas through the mesh points near $\Phi = 1$.

9.   Is the difference between the consecutive iterates
          $f_{n-1}(\theta)$ and $f_n(\theta)$ smaller than a prescribed tolerance,
          say ε_c ? If so, then we have found our solution. If not,

then in view of the new shape of the solid/liquid interface we recalculate the coefficient $\beta^{(i)}$, estimate the
values of Φ at the new mesh points from interpolation
and return to step 2.

It should perhaps be pointed out that before equation
(2-8) is discretised the independent variables are transformed
following Dennis *et al.* [1981]. This modification allows the
use of central differences despite the relatively high value of
the Peclet number, but care is needed in formulating some of
the partial derivatives near the boundaries. Full details of
the complete algorithm can be found in Craine & Andrews [1984].

4.   RESULTS AND DISCUSSION

For all the numerical results presented in this section
we have put $Pr = 0.1$ and $q = 2$ (again appropriate values for
the welding of steel). Various parameters which control the
computing time and accuracy of our solution must also be specified. It has been found that the computing time is considerably
reduced by using larger values of the tolerance parameters in
the early stages of the programme in order to obtain an approximate position of the solid/liquid interface and then decreasing
these parameters in stages to gradually refine the process. A
three-stage procedure with final values $\varepsilon = \varepsilon_\Omega = \varepsilon_\psi = \varepsilon_\Phi = 10^{-4}$
and $\varepsilon_c = 10^{-3}$ and with the relaxation parameters in the two *SOR*
schemes both equal to 1.1 has been used to calculate all the
displayed results.

The profiles of the solid/liquid interface for four different values of the decay radius ($R_0 = 0.6$, 0.8, 1.0 and 1.2)
for a typical welding current of 100 A (corresponding to
$K = 2.3 \times 10^5$) are shown in Figure 4, which also shows the equivalent profiles for the corresponding distributed heat sources
when fluid motion is ignored. In both cases the pool deepens
as the decay radius decreases. When fluid motion is included
the calculated shapes of the solid/liquid interface increasingly

depart from the ones for a stagnant pool but they are in good qualitative agreement with the boundary shapes observed in practice.

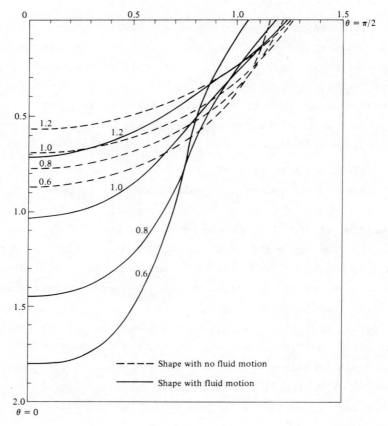

Figure 4

Weld pool shapes for various values of the decay radius R_0 = 0.6, 0.8, 1.0 and 1.2, for a current I = 100 A

Streamlines and isotherms inside the pool, again for a current of 100 A, are displayed in Figures 5a and 5b for R_0 = 1.2 and 0.6 respectively. The streamline patterns show single loop poloidal flows with fluid motion down the axis of symmetry. The isotherms are not significantly affected by the fluid motion in

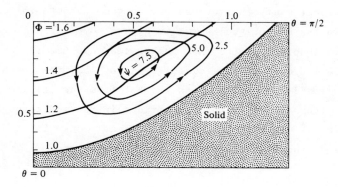

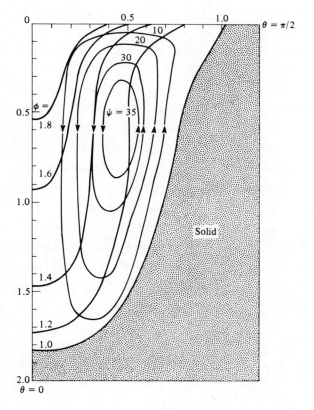

Figure 5
Streamlines and isotherms for a current $I = 100$ A
and decay radius $R_0 = 1.2$ and 0.6, respectively

the shallower pool of Figure 5a but they do tend to follow the motion in the deeper pool shown in Figure 5b.

It should be noted that the quoted results, all obtained with a 16×16 mesh in D_L, differ by about 15% from those found using a 11×11 mesh but by less than 3% from the solutions calculated with a 21×21 mesh. The errors introduced through our use of the 16×16 mesh are therefore greatly outweighed by the main trends shown by our results in Figures 4 and 5.

5.   CONCLUDING REMARKS

The results discussed in the previous section demonstrate that it is now possible to determine the shape of the weld pool when both the mechanisms of heat conduction and fluid flow are included and the stated numerical scheme has proved extremely robust for the cases considered. Undoubtedly more efficient numerical methods could be used but valuable qualitative information has been obtained from our results on the significant effect of electromagnetic stirring forces on the shape of the weld pool. Extensions of the basic model to include variations in surface tension, buoyancy, applied magnetic fields, moving sources, different workpiece geometry etc. all now seem within reach.

Following the first analytical solutions for a stagnant weld pool the advent of computers has enabled numerical solutions to be obtained for the more realistic models that take weld pool motion into account. The impact of these solutions on practical welding technology will be interesting to observe.

REFERENCES

Atthey, D.R. 1980, A mathematical model for fluid flow in a weld pool at high currents. *J. Fluid Mechs*, **98**, 787-801.

Carslaw, H.S. & Jaeger, J.C. 1959, *Conduction of Heat in Solids* (2nd edition). Clarendon Press, Oxford.

Craine, R.E. & Andrews, J.G. 1984, The shape of the fusion
    boundary in an electromagnetically stirred weld pool. Proc.
    IUTAM Symposium on *Metallurgical Applications of Magneto-
    hydrodynamics*, The Metals Society, 301-313.

Curle, N. 1962, *The Laminar Boundary Layer Equations*.
    Clarendon Press, Oxford.

Dennis, S.C.R., Ingham, D.B. & Singh, S.N. 1981, The steady
    flow of a viscous fluid due to a rotating sphere. *Q. J. Mech.
    Appl. Math.* **34**, 361-381.

Nestor, O.H. 1962, High intensity and current density distri-
    butions at the anode of high current, inert gas arcs.
    J. Appl. Phys. **33**, 1638-1648.

Sozou, C. 1971, On fluid motions induced by an electric current
    source. *J. Fluid Mech.* **46**, 25-32.

Thompson, J.F., Thames, F.C. & Martin, C.W. 1974, Automatic
    numerical generation of body-fitted curvilinear co-ordinate
    system for field containing any number of arbitrary two-
    dimensional bodies. *J. Comp. Phys.*, **15**, 299-319.

Woods, R.A. & Milner, D.R. 1971, Motion in the weld pool in arc
    welding, *Weld. J. Res. Suppl.* **50**, 1635-1715.

10

A numerical model of the glass sheet and fibre updraw processes

C. SAXELBY AND J. M. AITCHISON

1. INTRODUCTION

In this chapter we look at the numerical modelling of two processes in the glass industry. The first is the upward sheet drawing process, which, although presenting an interesting mathematical problem, is now rather dated and is no longer generally used for the commercial manufacture of glass sheet. This will be described by a model in two dimensional planar coordinates. The second process is the production of glass rods or fibres by the updraw process which is still of interest to the glass industry. Mathematically this is similar to, and an extension of, the first problem, being modelled in axisymmetric polar coordinates. The sheet updraw problem was first brought to our attention when it was presented at the first meeting of the University Consortium for Industrial Numerical Analysis by Pilkington Brothers PLC. The original object of the project was to model this process numerically and try to fit the results with the few analytic results that they already had (see Section 5). The glass fibre problem arose as a natural extension of the two-dimensional work.

Since the numerical modelling of flows of this kind is notoriously difficult, we will consider the two-dimensional problem in detail first. This has two advantages: firstly, there is much experience in the glass industry of the behaviour of this

199

process and secondly any lessons learnt here will be applied later to the more difficult axisymmetric problem

2.   THE TWO-DIMENSIONAL SHEET UPDRAW PROBLEM

The sheet updraw process is illustrated in Figure 1. Rollers in a tower draw up a sheet from a pool of molten glass. Rolls at the edge prevent it from 'necking in' and so variations across the width of the sheet can be neglected. Thus we can consider the situation shown in Figure 1 as a cross-section with the sheet of glass normal to the plane of the paper, giving a problem in two dimensions.

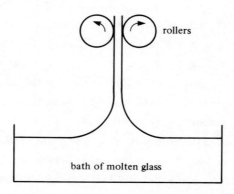

Figure 1

The glass is modelled as a Newtonian viscous fluid with a Reynolds number which is sufficiently small that the inertia terms can be neglected. The viscosity, in fact, varies rapidly with the temperature of the glass, and it is desirable that ultimately the problem should be solved to determine the temperature distribution. However, here we consider an isothermal model since this gives some indication of the nature of the solution and the feasibility of using a numerical method to calculate such a flow.

The major difficulty in calculating the fluid flow is that the shape of the cross-section in Figure 1 is unknown and has to be found as part of the solution. The glass/air interface is known as the free boundary and its position will determine the shape of the cross-section.

The governing differential equations can be solved most conveniently using a finite element method since much of the boundary is curved and will be moved during the iteration for the free boundary position. Thus it is necessary to consider a finite flow region and we consider the model shown in Figure 2 where we have made use of symmetry and show only half of the flow pattern. The glass sheet has been cut off at a height H where an outlet condition will be applied. The half-width of the tank is w, which is finite. The free surface is the line AB, the position of which is unknown. As it stands we have a moving boundary problem since the level of molten glass in the bath will drop as the sheet is drawn out. This is not convenient numerically nor does it reflect the physical situation. Therefore the problem is converted into one of steady flow by imposing some compensating fluid input across BC.

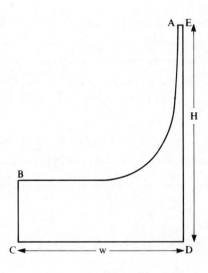

Figure 2

3.   THE MATHEMATICAL MODEL

The governing differential equations are the Navier Stokes equations without the inertia terms:

$$\frac{\partial}{\partial x}\left(t_{xx}\right) + \frac{\partial}{\partial y}\left(t_{xy}\right) = 0 \tag{3-1}$$

$$\frac{\partial}{\partial x}\left(t_{yx}\right) + \frac{\partial}{\partial y}\left(t_{yy}\right) - \rho g = 0 \tag{3-2}$$

and the equation of continuity

$$\frac{\partial u}{\partial x} + \frac{\partial v}{\partial y} = 0 \tag{3-3}$$

where u, v are horizontal and vertical components of velocity

ρ is the density (constant)

g is the acceleration due to gravity

and the stresses t_{xx}, t_{xy}, t_{yx}, t_{yy} are defined by

$$t_{xx} = -p + 2\mu\,\frac{\partial u}{\partial x}$$

$$t_{xy} = t_{yx} = \mu\left(\frac{\partial u}{\partial y} + \frac{\partial v}{\partial x}\right)$$

$$t_{yy} = -p + 2\mu\,\frac{\partial v}{\partial y}$$

where p is the pressure, and μ is the viscosity (constant). The solution of these equations on a fixed domain would normally require two boundary conditions on each section of the boundary. However, here we have the added difficulty of the unknown position of the free surface AB and we will find that an 'extra' boundary condition is needed there.

Consider a particular section of the boundary with outward normal n. Then the surface tractions are defined by

$$T_x = t_{xx}\,\frac{\partial x}{\partial n} + t_{xy}\,\frac{\partial y}{\partial n}$$

$$T_y = t_{yx}\,\frac{\partial x}{\partial n} + t_{yy}\,\frac{\partial y}{\partial n}.$$

The boundary conditions can now be written as follows:

On DE (axis of symmetry): $u = 0$, $T_y = 0$. $\tag{3-4}$

On *EA* (outlet) $T_x = 0,\ v = V_0$ (given)   (3-5)

On *AB* (free surface) $T_x = T_y = 0$,

$$u\,dy - v\,dx = 0 \ . \tag{3-6}$$

(streamline condition)

The imposition of a suitable inlet condition to produce a steady flow proved to be more difficult than anticipated. After much trial and error we found two alternative sets of inlet conditions which gave satisfactory results.

Inlet 1

On *BC*: $T_x = p - 2\mu\ \dfrac{\partial u}{\partial x} = \rho g(s-y)$ (3-7a)

$v = 0$.

On *CD* (no-slip base) $u = v = 0$.   (3-8a)

Inlet 2

On *BC*: $u = U_0$ (given), $v = 0$ (3-7b)

On *CD*: $T_x = 0,\ v = 0$.   (3-8b)

For both cases the boundary conditions are divided into two groups as follows. We *specify* the outlet height, H, width, h, and velocity, V_0. We also specify the position of the inlet, w. We then *calculate* the inlet height *BC* and the shape of the free surface *AB* .

It is clear that we cannot specify everything at both the inlet and outlet in advance. The above strategy was arrived at after much experimentation. In particular it was found that fixing the inlet height and allowing the outlet width or height to vary did not produce a convergent numerical scheme. This will be discussed in more detail later.

4.   NUMERICAL SOLUTION

The solution process involves two main steps — the solution of the differential equations and the location of the free

surface.   The free surface position is found iteratively as
follows:

(1)   Guess a position for the free boundary, AB, giving a
         flow region R.
(2)   Solve the differential equations using the stress condi-
         tions on the free surface.
(3)   Move the free surface to satisfy the streamline condi-
         tion — defining a new region R.

         Steps (2) and (3) are repeated until a free surface is
found on which the stress and streamline conditions are satisfied simultaneously.   Thus we have to solve the differential
equations on a sequence of similar regions.   This is done using
the Galerkin form of the finite element method (see, for example,
Mitchell & Wait [1977] or Zienkiewicz [1977]).   The region R is
divided into triangular elements based on a radial mesh, as shown
in Figure 3.   As the region R changes from one iteration to the
next the elements are distorted to fit the new boundary shape.
The number of elements and the linkage structure between them
remains constant as the boundary is moved.   This is important in

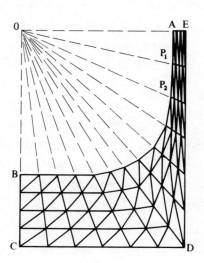

Figure   3

the solution of the sequence of finite element equations since many of the time consuming 'house-keeping' operations need only to be done once. These include the setting up of lists relating the nodes to elements which will be needed in the matrix assembly and the calculation of a sparseness structure and pivotal strategy for the solution of the algebraic system. Neither of these depend on the absolute values of the nodal coordinates — they depend on the topology of the grid layout rather than on its geometry.

Following Nickell *et al.* [1974] we consider the weak form of the differential equations in stress form, where the stress conditions arise as natural boundary conditions.

Let

$$u \simeq \tilde{u} = \sum_{i=1}^{N} u_i \, \phi_i \, (x,y)$$

$$v \simeq \tilde{v} = \sum_{i=1}^{N} v_i \, \phi_i \, (x,y)$$

and

$$p \simeq \tilde{p} = \sum_{i=1}^{M} p_i \, \psi_i \, (x,y)$$

where

$\phi_i (x,y)$, $i = 1,\ldots,N$ and $\psi_i(x,y)$, $i = 1,\ldots,M$ are given basis functions. In general $\phi_i(x,y)$ and $\psi_i(x,y)$ will be different and the set of nodes associated with each set of basis functions may or may not overlap. Let R be the region occupied by the fluid then the functions $\tilde{u}$, $\tilde{v}$ and $\tilde{p}$ are chosen to satisfy the weak forms of equations $(3-1)-(3-3)$ as follows:

$$\int_R \left\{ -\frac{\partial}{\partial x}(\tilde{t}_{xx}) - \frac{\partial}{\partial y}(\tilde{t}_{xy}) \right\} \phi_i \; dx \, dy = 0, \quad i = 1,\ldots,N \quad (4\text{-}1)$$

$$\int_R \left\{ \rho g - \frac{\partial}{\partial x}(\tilde{t}_{yx}) - \frac{\partial}{\partial y}(\tilde{t}_{yy}) \right\} \phi_i \, dx \, dy = 0, \quad i = 1,\ldots,N \quad (4\text{-}2)$$

and

$$\int_R \left\{ \frac{\partial \tilde{u}}{\partial x} + \frac{\partial \tilde{v}}{\partial y} \right\} \psi_i \, dx \, dy . \qquad\qquad i = 1,\ldots,M \quad (4\text{-}3)$$

These form $(2N + M)$ algebraic equations for the coefficients

$$u_i, v_i, \quad i = 1, \ldots, N \quad \text{and} \quad p_i, \quad i = 1, \ldots, M.$$

If a particular u_i or v_i is given as a boundary condition then the corresponding equation is deleted from the set. Note that equation (4-3) uses the basis function ψ_i associated with the pressure, although the corresponding differential equation, the continuity equation, does not involve p directly.

Applying the divergence theorem to equation (4-1) we obtain

$$\int_R \left\{ \tilde{t}_{xx} \frac{\partial \phi_i}{\partial x} + \tilde{t}_{xy} \frac{\partial \phi_i}{\partial y} \right\} dx\, dy - \int_S T_x \phi_i\, ds = 0\,, \quad i = 1, \ldots, N \tag{4-4}$$

where S is the boundary of the region R and T_x is the x component of the surface traction defined earlier. This surface integral makes the imposition of boundary conditions straightforward: either u_i is given on S and then this particular equation does not arise (since u_i is not free), or T_x is prescribed. In particular the condition $T_x = 0$ is a 'natural boundary condition' for this equation.

Similarly, equation (4-2) can be rewritten as

$$\int_R \left\{ \rho g \phi_i + \tilde{t}_{yx} \frac{\partial \phi_i}{\partial x} + \tilde{t}_{yy} \frac{\partial \phi_i}{\partial y} \right\} dx\, dy - \int_S T_y \phi_i\, ds = 0\,,$$
$$i = 1, \ldots, N \tag{4-5}$$

where the natural boundary condition is now $T_y = 0$.

Thus we solve equations (4-3) $-$ (4-5) for the coefficients $\{u_i\}$, $\{v_i\}$ and $\{p_i\}$.

These equations involve first derivatives of u and v, and only function values of p. Therefore the minimum requirements for the basis functions ϕ_i and ψ_i are

(i) ϕ_i : linear in each element, $C^{(0)}$ continuity between elements.

(ii) ψ_i : constant in each element, no continuity required between elements.

These basis functions do not give a very accurate solution and
so for the present work we use the popular approximations of
piecewise quadratic functions for u and v and linear functions
for p, both of which have $C^{(0)}$ continuity between the elements.
This combination avoids any oscillations in the solution for the
pressure (Jackson & Cliffe [1980]).

The resulting algebraic equations involve a matrix which
is sparse but not symmetric. Furthermore many of the rows have
zero diagonal elements. Therefore the equations are solved using
a sparse form of Gaussian elimination with pivoting.

When the values of u, v and p have been calculated for
a given region R, the boundary AB must be moved in an attempt
to satisfy the streamline condition. This is achieved by moving
each boundary node along the 'radial' lines shown dashed in
Figure 3. Starting from the fixed point A, we calculate new
boundary nodes P_1, P_2, ..., so that gradient of $P_{k-1} P_k$

$$= \frac{\displaystyle\int_{P_{k-1}}^{P_k} v\,ds}{\displaystyle\int_{P_{k-1}}^{P_k} u\,ds}$$

using the calculated values of u and v. The finite element
grid is then stretched and the whole process repeated.

5.   RESULTS

The techniques described in the previous section were
first applied to the more standard problem of extrusion of a
viscous fluid from a pipe. This problem is steadily acquiring
the status of a test problem for codes for the calculation of
viscous free surface flows. The results obtained from the
present algorithm agree well with those of Nickell et al. [1974],
Chang et al. [1979] and Omodei [1980]. These results are all
obtained using a finite element approach but with a variety of
basis functions. The major object behind this comparison is

C. SAXELBY AND J.M. AITCHISON

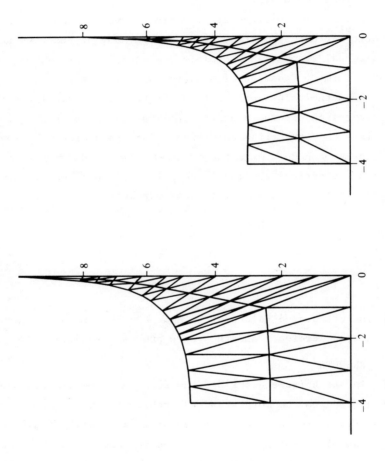

Figure 4a,b

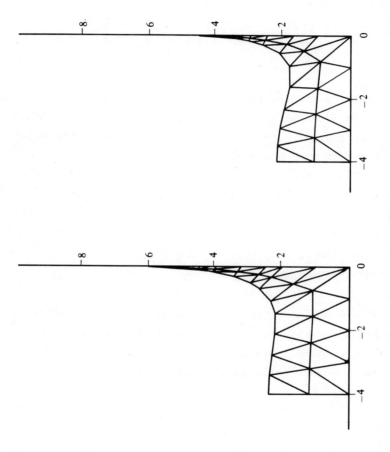

Figure 4c,d

merely to test the program developed here. These results are
discussed in Saxelby [1981].

For the present problem, calculations were performed for
a variety of the parameters describing the geometry of the process. Typical results using inlet condition 1 are shown in
Figures 4a, b, c, d. For each of these flow-patterns we have
taken

$$V_0 = 40 , \quad h = 0.025 , \quad w = 4 .$$

$$\mu = \rho = g = 1 .$$

The values of H vary from 10 to 4.5 as indicated in the dia-
grams. We see that the shape of the free surface is basically
the same in each case but that there is a zone of adjustment
near the left-hand boundary.

This phenomenon is even more noticeable when we use the
second set of inlet conditions. Figures 5a, b, c, illustrate
typical free surface shapes obtained by using various values of
the prescribed inlet velocity U_0. These results were all obtained for the same value of H and can be virtually superimposed
for most of the flow region.

It is important to make some comments on the convergence
of the free surface iteration. In the case of the simple extrusion problem where the free surface is fixed at one end we found
that the boundary iteration described here worked very well and
that the boundary position had effectively converged after 4 or

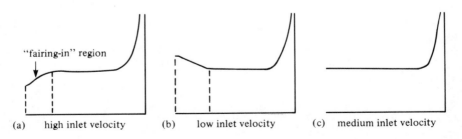

(a)   high inlet velocity   (b)   low inlet velocity   (c)   medium inlet velocity

Figure 5a, b, c

5 iterations. Thus the iteration converges much faster than the corresponding procedure for potential problems (see, e.g., Aitchison [1972]). This is due to the stable nature of the viscous flow. However, the use of the streamline condition as the 'moving condition' can present problems if the stream function itself is not one of the primary variables. We saw, in the previous section, that the moving algorithm is based on the calculation of a gradient and a projection forward to the next boundary point. In the case of the glass updraw problem we have some freedom in the choice of a fixed point from which to start the boundary movement (inlet or outlet). After many failures we found that the best strategy was to fix the point A at the outlet and move down the free boundary against the direction of flow. We also found that only certain combinations of inlet and outlet boundary conditions allowed a successful boundary iteration. The one redeeming feature of this problem is that convergence is rapid when it occurs, and therefore non-convergence is easily recognized.

One of the original reasons for looking at the two dimensional problem was to test the prediction of Gelder [1977] who used a lubrication theory approach to examine the shape of the free surface near the outlet. This was shown to behave like

$$h = \frac{8\mu Q}{g l^2 \cosh^2 (y/l+b)}$$

where h is the half-width of the sheet, Q is the total flow and l and b are constants to be determined.

This curve was fitted to the numerical results obtained here, and a good fit obtained in the upper region of the free surface. This fitting emphasized the fact that as the outlet length is increased the free surface shape changes very little.

6.   THE GLASS FIBRE PROBLEM

In this process a fibre is drawn up from a pool of molten glass in an axisymmetric fashion. The process can again be described by Figure 2, provided that the axes are taken to be r and z. If we now let u and v be the fluid velocities in the radial and axial directions then the governing equations become

$$\frac{\partial}{\partial r}(t_{rr}) + \frac{\partial}{\partial z}(t_{rz}) + \frac{2\mu}{r}\left(\frac{\partial u}{\partial r} - \frac{u}{r}\right) = 0 \ , \tag{6-1}$$

$$\frac{\partial}{\partial r}(t_{zr}) + \frac{\partial}{\partial z}(t_{zz}) + \frac{2\mu}{r}\left(\frac{\partial v}{\partial r} + \frac{\partial u}{\partial z}\right) - \rho g = 0 \ , \tag{6-2}$$

$$\frac{\partial u}{\partial r} + \frac{\partial v}{\partial z} + \frac{u}{r} = 0 \ , \tag{6-3}$$

where

$$t_{rr} = -p + 2\mu\frac{\partial u}{\partial r}$$

$$t_{rz} = t_{zr} = \mu\left(\frac{\partial u}{\partial z} + \frac{\partial v}{\partial r}\right)$$

$$t_{zz} = -p + 2\mu\frac{\partial v}{\partial z} \ .$$

These equations are very similar to equations $(3-1) - (3-3)$ for the planar problem but contain extra terms which may cause problems when r is close to zero. We apply the Galerkin technique as before, but derive the weak forms of the differential equations by multiplying by r and then integrating with respect to r and z. This is equivalent to integrating over the whole of the three dimensional physical flow region, and has the mathematical attraction of removing most of the $1/r$ terms which would otherwise arise.

After applying the divergence theorem we obtain

$$\int_R \left\{ t_{rr}\frac{\partial}{\partial r}(r\phi_i) + t_{rz}\frac{\partial}{\partial z}(r\phi_i) - 2\mu\frac{\partial u}{\partial r}\phi_i \right.$$

$$\left. + 2\mu\frac{u}{r}\phi_i \right\} dr\, dz - \int_S T_r r\phi_i\, ds = 0 \ , \tag{6-4}$$

$$\int_R \left\{ t_{zr} \frac{\partial}{\partial r} (r \phi_i) + t_{zz} \frac{\partial}{\partial z} (r \phi_i) - \mu \left(\frac{\partial v}{\partial r} + \frac{\partial u}{\partial z} \right) \phi_i \right.$$
$$\left. + r g \phi_i \right\} dr \, dz - \int_S T_z \, r \phi_i \, ds = 0 , \qquad (6\text{-}5)$$

and

$$\int_R \left\{ r \frac{\partial u}{\partial r} + r \frac{\partial v}{\partial z} + u \right\} \psi_i \, dr \, dz = 0 , \qquad (6\text{-}6)$$

as the $(2N+M)$ equations to be solved. Note that the natural boundary conditions for equations (6-4) and (6-5) are now $r T_r = 0$ and $r T_z = 0$ respectively. This does not present any problems on the free boundary where $r \neq 0$, but the condition on the axis of symmetry becomes $r T_z = 0$ on $r = 0$. However this seems to work well and is used throughout.

The inlet boundary conditions used for the fibre problem were the same as those described earlier for the planar problem. However it was much more difficult to obtain convergent itera- tions for the free boundary position for this problem, and con- verged solutions were only obtained for a limited range of geometries. Part of the problem of convergence may be due to the wide range of velocities involved. Certainly the easiest geometries were those with small width or giving a small flux and consequently a relatively small outlet velocity. However the width cannot be made too small without causing problems with the outlet condition. The best results were obtained between these extremes.

7.   CONCLUSIONS

We have described our attempts to formulate a numerical model of the glass updraw problem. The major difficulties which arise are due firstly to the combination of the viscous flow equations and the unknown free surface and secondly, to the need to impose artificial inlet and outlet conditions. It is this second problem which has made the development of a convergent numerical scheme difficult. The results given here indicate at

least a partial success, but it seems that there are still problems near the left hand end of the flow and that the present sets
of inlet conditions, although adequate, are not ideal. It should
be noted that taking entirely natural boundary conditions at the
inlet, which is very attractive in theory, does not work in practice since the boundary iteration does not converge. In summary,
it seems that too much freedom at the inlet leads to a lack of
convergence of the free boundary iteration and that the more
specific inlet conditions given in equations (3-7) and (3-8) give
rise to a convergent boundary iteration, but to a boundary shape
whose left hand end does not match comfortably with the right.
Perhaps a more careful choice of inlet conditions will overcome
this difficulty.

It would obviously be of great interest to the glass industry to have a numerical model of the glass fibre process which
not only included the effects described here but also those of
surface tension and the temperature dependent viscosity of the
glass. The numerical models described here fall a long way short
of this target. However, they are a substantial improvement on
the previous attempt by Gelder [1977] to model these processes
using lubrication theory. The programs developed here are now
being used by Pilkingtons as research tools, but there is
obviously scope for further work.

ACKNOWLEDGEMENTS

The problem of glass updraw was first brought to the
authors' attention by Mr D. Gelder of Pilkington Brothers PLC,
when he presented it to the first meeting of the University
Consortium for Industrial Numerical Analysis in January 1980. We
are very grateful to him for many helpful discussions. We are
also grateful to the Science and Engineering Research Council
who supported the work of C. Saxelby through a Research Student-
ship.

REFERENCES

Aitchison, J.M., 1972 Numerical treatment of a singularity in a
free boundary problem, *Proc. Roy. Soc. Lond. A* **330**, 573-580.

Chang, P-W, Patten T.W. and Finlayson, B.A. 1979 Collocation
and Galerkin Finite Element Methods for viscoelastic fluid
flow — II, *Comp. Fluids*, **7**, 285-294.

Gelder, D. 1977 The basis of upward sheet drawing, *Glass
Technology*, **18**, 178-180.

Jackson, C.P. and Cliffe, K.A. 1980 Mixed interpolation in.
primitive variable finite element formulations for incompressible flow, AERE Harwell Report No. T.P.866.

Mitchell, A.R. and Wait, R. 1977 *The Finite Element Method in
Partial Differential Equations*, Wiley.

Nickell, R.E., Tanner. R.I. and Caswell. B. 1974 The solution
of viscous incompressible jet and free surface flows using
finite element methods, *J. Fluid Mech*, **65**, 189-206.

Omodei, B. 1980 On the dieswell of an axisymmetric Newtonian
jet, *Comp. Fluids*, **8**, 275-289.

Saxelby, C. 1981 The solution of partial differential equations by the finite element. D.Phil thesis, University
of Oxford.

Zienkiewicz, O.C. 1977 *The Finite Element Method* (3rd edition)
McGraw Hill.

11

The motion of a towed array of hydrophones

R. CARTWRIGHT AND D. F. MAYERS

1.   INTRODUCTION

This problem, originally raised by Plessey Marine Research, is concerned with the motion of an array of hydrophones towed behind a ship at sea.   A hydrophone is a device for detecting sound waves under water;   simultaneous readings from an array of hydrophones make it possible to determine the position of the source of sound waves.   However, it is necessary to know quite accurately the positions of the hydrophones.   The array is contained in a slightly flexible cylinder several kilometers long, and is towed behind a ship. Under ideal conditions the sea would be at rest and the ship would move with uniform velocity in a straight line; the hydrophones would then lie in a straight line along the direction of motion.   In practice there will be weather disturbances affecting the motion of the ship and ocean currents disturbing the cylinder. The object of this study is to discover what effect these disturbances have in a simple case. The simple model problem proposed by Plessey was to impose on the uniform motion of the ship a periodic transverse motion and to determine the subsequent motion of the cylinder.   The particular quantity of interest is the curvature of the cylinder which measures its departure from a straight line.

In the next section we derive a hyperbolic partial

differential equation which describes the motion of the cylinder. In Sections 3 and 4 we describe two numerical methods used for the solution of this equation. The final section presents specimen results.

2. MATHEMATICAL MODEL

The forces on a submerged thin flexible cylinder are well established; the same physical problem arises in the study of an anchor cable in a flowing stream, and of the umbilical system connected to a submersible of the kind used in underwater exploration. The mathematical model employed here is derived from the treatment given by Ferriss [1981] and by Huffman & Genin [1971].

The array which we are considering can also be treated as a thin flexible cylinder similar to a cable. It is arranged to be neutrally buoyant and floats just below the surface of the sea. The motion is therefore confined to two dimensions and gravity forces can be neglected. We use the following notation in suitably scaled units:

s arc length along the cable;

u, w tangential and normal components of the velocity of the sea, relative to the cable;

ψ the angle between the cable and the flow direction;

T the tension in the cable.

The first two equations are the standard kinematical conditions, relating the motion of the cable to the angle ψ,

$$\frac{\partial u}{\partial s} = - \varepsilon^2 \, w \, \frac{\partial \psi}{\partial s} \qquad (2\text{-}1)$$

$$\frac{\partial w}{\partial s} = u \, \frac{\partial \psi}{\partial s} + \frac{\partial \psi}{\partial t} \; . \qquad (2\text{-}2)$$

The momentum equations, for motion in the tangential and normal directions, are:

$$\frac{\partial T}{\partial s} = - \tfrac{1}{2} \, Q u - \alpha \, \frac{\partial u}{\partial t} - \varepsilon^2 \, \alpha w \, \frac{\partial \psi}{\partial t} \qquad (2\text{-}3)$$

$$(T - \alpha u^2) \frac{\partial \psi}{\partial s} - \frac{1}{2} w (Q + 2\beta |w|) = 2\alpha \frac{\partial w}{\partial t} \qquad (2\text{-}4)$$

where $Q^2 = u^2 + \varepsilon^2 w^2$ and α, β, and ε are non-dimensional
parameters. In these equations the terms $\frac{1}{2} Q u$ and $\frac{1}{2} w (Q + 2\beta |w|)$
represent the components of the drag force exerted by the sea.
The term in $u \frac{\partial \psi}{\partial t}$ which naturally arises in (2-4) has been combined with the other terms by using the second kinematic equation
(2-2).

Although the parameter α is quite small, it cannot be
neglected in equation (2-4) because the tension T becomes small
near the free end of the cable and αu^2 is comparable to T. However, the parameter ε^2 is negligible; ε has the value

$$\varepsilon = a/V\tau , \qquad (2\text{-}5)$$

where V is the forward speed of the ship, a is the amplitude
of the transverse motion of the ship and τ is the period of the
transverse motion. In the units used here we find that ε^2 is
about 10^{-4}, so the last term can be neglected in (2-3), and also
equation (2-1) shows that u is a function of t only. If we
assume that the ship maintains a steady forward speed we can
therefore choose $u = Q = 1$.

Equation (2-3) now becomes

$$\frac{\partial T}{\partial s} = -\frac{1}{2} \qquad (2\text{-}6)$$

so that

$$T(s,t) = -\frac{1}{2}(s - s_1) + T_1(t)$$

where $T_1(t)$ is the tension at the free end of the cable, the
point $s = s_1$. The value of $T_1(t)$ is determined by the boundary conditions imposed at the free end; as we shall see later,
these boundary conditions give a constant value to $T_1(t)$, so
that T_1, and hence also T, is independent of t.

Equation (2-2) becomes, writing $u = 1$,

$$\frac{\partial}{\partial s} (w - \psi) = \frac{\partial \psi}{\partial t} , \qquad (2\text{-}7)$$

so that we can write w and ψ in terms of a function y,

$$\psi = \frac{\partial y}{\partial s} \ , \qquad w - \psi = \frac{\partial y}{\partial t} \tag{2-8}$$

and

$$w = \frac{\partial y}{\partial s} + \frac{\partial y}{\partial t} \ . \tag{2-9}$$

This new function y is closely related to the displacement of the cylinder from the straight position. If this displacement, $\bar{y}$, were very small, we should have $\frac{\partial \bar{y}}{\partial s} = \sin\psi \sim \psi$, but in any case we have

$$\frac{\partial^2 y}{\partial s^2} = \frac{\partial \psi}{\partial s}$$

so that $\partial^2 y / \partial s^2$ is the curvature of the cylinder.

Rewriting equation (2-4) in terms of the new function $y(s,t)$, using the usual subscript notation to denote differentiation, gives:

$$(T - \alpha) y_{ss} - \tfrac{1}{2}(y_s + y_t)\left\{ 1 + 2\beta | y_s + y_t | \right\} = 2\alpha(y_{st} + y_{tt}) . \tag{2-10}$$

To complete the specification of the problem we must provide boundary and initial conditions. At the leading end of the cylinder the motion is prescribed by the motion of the ship; this gives a transverse simple harmonic motion, and it is convenient to use the condition

$$y(0,t) = 1 - \cos(\lambda t) , \tag{2-11}$$

where λ is a prescribed frequency. At the trailing end, where $s = s_1$, we assume that there is zero crossflow, so that $w = 0$, or

$$y_s + y_t = 0 \quad \text{at} \quad s = s_1 . \tag{2-12}$$

It is clear from equation (2-4) that if $T - \alpha u^2$ becomes negative the motion becomes unstable, giving rise to a whiplash effect. To stop this happening, some form of drogue is attached to the end of the cylinder; this is a very simple device which produces a given drag in the water, so that the tension in the cylinder at the free end can be assumed to be a given positive constant. We can then write

$$T - \alpha = -\frac{1}{2}(s - s_1) + T_1 - \alpha$$

$$= (A - s)/2 \tag{2-13}$$

where A is a given constant, with $A > s_1$.

We suppose that initially the cylinder is straight, with no transverse velocity, so that $y = 0$ and $y_t = 0$.

The numerical problem is now to find the solution of the second order hyperbolic equation

$$(A - s)y_{ss} - 4\alpha y_{st} - 4\alpha y_{tt} - (y_s + y_t)(1 + 2\beta|y_s + y_t|) = 0 \tag{2-14}$$

on $0 < s < s_1$, $t > 0$, with the boundary conditions

$$y(s,0) = 0, \quad y_t(s,0) = 0,$$
$$y(0,t) = 1 - \cos(\lambda t), \quad y_s(s_1,t) + y_t(s_1,t) = 0. \tag{2-15}$$

Typical values of the constant parameters are

$$\lambda = 1, \quad A = 0.35, \quad s_1 = 0.25, \quad \alpha = 1/30, \quad \beta = 3. \tag{2-16}$$

3.   FINITE DIFFERENCE SOLUTION

We use a uniform grid with a spatial step $h = s_1/N$, where N is an integer, and a time step k. Writing y_{ij} as usual for the numerical approximation to $y(ih, jk)$ the standard central difference approximations for y_s, y_t, y_{ss} and y_{tt} are

$$y_s \doteq \frac{(\delta_s y)_{ij}}{2h} = \frac{1}{2h}(y_{i+1,j} - y_{i-1,j})$$

$$y_t \doteq \frac{(\delta_t y)_{ij}}{2k} = \frac{1}{2k}(y_{i,j+1} - y_{i,j-1})$$

$$y_{ss} \doteq \frac{(\delta_s^2 y)_{ij}}{h^2} = \frac{1}{h^2}(y_{i+1,j} - 2y_{ij} + y_{i-1,j}) \tag{3-1}$$

$$y_{tt} \doteq \frac{(\delta_t^2 y)_{ij}}{k^2} = \frac{1}{k^2}(y_{i,j+1} - y_{ij} + y_{i,j-1}).$$

The cross derivative may be approximated by

$$y_{st} \doteq \frac{1}{4hk}(y_{i+1,j+1} - y_{i+1,j-1} - y_{i-1,j+1} + y_{i-1,j-1}).$$

$$(3\text{-}2)$$

Substituting these approximations into the differential equation (2-14) we shall evidently obtain an implicit equation, as it will involve three unknowns $y_{i-1,j+1}$, $y_{i,j+1}$ and $y_{i+1,j+1}$ on the new time step, $j+1$; this is of course due to the presence of the cross derivative y_{st}. It would be possible to make the scheme explicit, and therefore computationally simpler, by using another difference approximation y_{st} which did not involve $y_{i-1,j+1}$ or $y_{i+1,j+1}$. However such approximations are significantly less accurate and in practice it has been found that the solution of the implicit equations causes no difficulty.

Having accepted that the scheme will be implicit, we can improve the stability of the solution without increasing the computational labour much further by using the approximations

$$y_{ss} \doteq \frac{1}{4h^2}(\delta_s^2 y_{i,j+1} + 2\delta_s^2 y_{ij} + \delta_s^2 y_{i,j-1}) \qquad (3\text{-}3)$$

$$y_s + y_t \doteq \frac{1}{8h}(\delta_s y_{i,j+1} + 2\delta_s y_{ij} + \delta_s y_{i,j-1}) + \frac{1}{2k}\delta_t y_{ij}.$$

$$(3\text{-}4)$$

A simple von Neumann stability analysis shows that use of (3-1) requires the condition $k/h \leq (4\alpha/A)^{\frac{1}{2}}$, whereas use of (3-3) and (3-4) is unconditionally stable.

The finite difference equation then involves nine neighbouring values, with subscripts, $i-1$, i, $i+1$ and $j-1$, j, $j+1$. Of these the six with subscripts $j-1$ or j are known, and the equation for the unknown values becomes

$$\tfrac{1}{4}(A-ih)(z_{i-1} - 2z_i + z_{i+1})/h^2$$

$$-4\alpha(z_{i+1} - z_{i-1})/4hk - 4\alpha z_i/k^2$$

$$-\left[\tfrac{1}{4}(z_{i+1} - z_{i-1})/2h + f_i\right]\left[1 + 2\beta|\tfrac{1}{4}(z_{i+1} - z_{i-1})/2h + f_i|\right] = g_i$$

$$i = 1, 2, \ldots, N-1, \qquad (3\text{-}5)$$

where z_i is the unknown $y_{i,j+1}$ and f_i, g_i represent known quantities available from previous time steps. In addition the boundary condition at $s = s_1$ leads to

$$\frac{z_N - y_{N,j}}{k} + \frac{y_{N,j} - y_{N-1,j}}{h} = 0 \qquad (3\text{-}6)$$

so that z_N can be eliminated from the last of equations (3-5), with $i = N-1$. The result is a set of simultaneous equations for the unknowns $z_1, z_2, \ldots, z_{N-1}$.

These equations are nonlinear, but Newton's method has been found to produce the required solution without difficulty. The Jacobian matrix required in the Newton iteration is evidently an $N-1 \times N-1$ tridiagonal matrix, and the calculation can easily be made very efficient. The iterative method of course requires an initial estimate of all the unknowns z_i. The fact that $\beta = 3$ suggests that the term in the differential equation (2-14) which involves 2β is likely to be the most important, and a first estimate of z_i is obtained by solving $y_s + y_t = 0$. This means that in equation (3-5) we simply find the solution of the system of linear equations

$$\frac{1}{4}(z_{i+1} - z_{i-1})/2h + f_i = 0 \qquad (3\text{-}7)$$

along with the boundary condition (3-6). This is a very simple tridiagonal system, and has been found to give a sufficiently close first estimate.

The incorporation of the boundary condition at $s = 0$ is trivial, since it just shows that $y_{0,j+1}$ is a known quantity. The initial conditions $y(s,0) = 0$ and $y_t(s,0) = 0$ are dealt with by starting the finite difference equations from $y_{i0} = 0$ and $y_{i,-1} = 0$ for $i = 1,2,\ldots,N-1$. Then the finite difference solution begins with $j = 1$, proceeding with $j = 2,3,\ldots$ at each stage solving the resulting system of equations (3-5) by Newton's method.

4.   THE METHOD OF CHARACTERISTICS

The method of characteristics is normally regarded as essential for the solution of hyperbolic equations involving singularities, since these singularities propagate through the region of interest. This problem has several weak singularities. The presence of the term $|y_s + y_t|$ in equation (2-14) shows that at every point at which $y_s + y_t$ passes through zero and changes sign the fourth derivatives of y are discontinuous; this singularity then propagates forward along the two characteristics through that point. The usual analysis of local truncation error therefore breaks down at points near these characteristics, since it assumes the existence of a continuous fourth derivative, and the error is therefore likely to be larger at these points than elsewhere. But this is a weak discontinuity in a high order derivative, and its effect is quickly damped by the first derivative terms in the equations; in numerical experiments it is found to have only a small effect on the accuracy of the solution.

There are also weak singularities arising from the corners of the region in the (s,t) plane; for example, on $s = 0$, $y = 1 - \cos(\lambda t)$, and on $t = 0$, $y = y_t = 0$. Successive differentiation of these conditions, and of equation (2-14), shows that as the point (s,t) tends to the origin along the two coordinate axes, the third derivative y_{stt} tends to different limits. Hence the third derivative is discontinuous at the origin, and therefore also at all points on the characteristic through the origin, This singularity is caused by the sudden initial motion of the ship, and is also quickly damped out.

These singularities therefore seem unlikely to have much effect on the accuracy of the finite difference solution, but it is still interesting to investigate the performance of the method of characteristics on this practical problem.

The characteristics of the equation (2-14) are the solutions of the ordinary differential equation

$$(A - s)\left(\frac{dt}{ds}\right)^2 + 4\alpha\left(\frac{dt}{ds} - 1\right) = 0 \qquad (4\text{-}1)$$

and along a characteristic the relation

$$(A - s)\frac{dt}{ds}\, dp - f\, dt - 4\alpha\, dq = 0 \qquad (4\text{-}2)$$

holds, where we have used the standard notation $p = y_s$, $q = y_t$, and we have also written

$$f = f(p,q) = (p+q)[1 + 2\beta|p+q|]. \qquad (4\text{-}3)$$

It is a simple matter to solve the equation for the characteristics, the result being

$$t \pm u + 4\alpha\, \log(u \pm 4\alpha) = \text{constant} \qquad (4\text{-}4)$$

where

$$u^2 = 16\,\alpha^2 + 16\,\alpha\,(A - s); \qquad (4\text{-}5)$$

the choice of signs giving the two systems of characteristics.

A numerical solution is then obtained by the usual method, on the same uniform grid of points as for the finite difference method. To obtain the solution at the point C, (s_i, t_{j+1}), we first find the two characteristics through this point, and then calculate the coordinates of the points A and B where these two curves cut the previous line $t = t_j$. For example, for the positive characteristic we must solve

$$t_{j+1} + u_C + 4\alpha\,\log(u_C + 4\alpha) = t_j + u_A + 4\alpha\,\log(u_A + 4\alpha) \qquad (4\text{-}6)$$

where t_j, t_{j+1} and u_C are known. This equation is easily solved for u_A by Newton's method, and then s_A is found, knowing u, from (4-5). The value of s_B is calculated by exactly the same method, but using the negative sign in (4-4).

Knowing the values of u, p and q at all points on the previous time step, $t = t_j$, we then calculate the values of u, p and q at the two points A and B by an interpolation process using values at the neighbouring grid points. Along the characteristics we then approximate (4-2) by

$$\tfrac{1}{2}\left\{(A-s_A)\left(\frac{dt}{ds}\right)_A + (A-s_C)\left(\frac{dt}{ds}\right)_C\right\}(p_C - p_A)$$

$$-\tfrac{1}{2}\left\{f(p_C,q_C) + f(p_A,q_A)\right\}k - 4\alpha(q_C - q_A) = 0 \qquad (4\text{-}7)$$

with a similar equation along the other characteristic. These are then two equations for the two unknowns p_C and q_C, the other quantities all being known; it will be found that if we subtract (4-7) from the other similar equation, we get a new equation involving p_C only, from which p_C can be found almost trivially. The value of q_C can then be obtained without difficulty from (4-7). Finally y_C is computed from the first terms of a Taylor series expansion in the form

$$y_C = y_A + \tfrac{1}{2}(s_C - s_A)(p_C + p_A) + \tfrac{1}{2}k(q_C + q_A). \qquad (4\text{-}8)$$

The method has to be modified in a fairly obvious way when the point C is on one of the boundaries. In this case we use only one of the characteristics, of course; on the left hand boundary y, and therefore y_t, is known, while on the right hand boundary we know that $y_s + y_t = 0$. These in each case give sufficient information to calculate the required unknowns.

In this description of the method we have solved the equations for the characteristics exactly, but used numerical approximations for the relations (4-2) along the characteristics. Without any significant loss of accuracy we can also use a numerical approximation to find the characteristics, thus obtaining numerical approximations to the values of s_A and s_B. A simple way of doing this is to solve the equations

$$k = t_C - t_A = \tfrac{1}{2}\left\{\left(\frac{dt}{ds}\right)_C + \left(\frac{dt}{ds}\right)_A\right\}(s_C - s_A)$$

$$\qquad\qquad\qquad\qquad\qquad\qquad\qquad\qquad (4\text{-}9)$$

$$k = t_C - t_B = \tfrac{1}{2}\left\{\left(\frac{dt}{ds}\right)_C + \left(\frac{dt}{ds}\right)_B\right\}(s_C - s_B)$$

for s_A and s_B, where

$$\frac{dt}{ds} = \frac{-4\alpha \pm \{16\alpha^2 + 16\alpha(A-S)\}^{\frac{1}{2}}}{2(A-s)} . \qquad (4\text{-}10)$$

The solution of these two equations then provides an alternative to the solution of equation (4-6) in the two cases. The final results are found to agree to within the required accuracy, and the use of equation (4-9) is significantly faster. This is an example of a situation where a numerical approximation gives the required accuracy, and is more efficient computationally, than the use of the known analytic solution for the characteristics.

5.   CONCLUSIONS

Using the same grid points, the numerical accuracy of the two methods has been found to be similar. Both give errors of the order of 10^{-3} with $N = 8$ and $k = 0.02$. The solution is very smooth, and is quite close to a straight line when $\lambda = 1$. If the frequency of the ship's motion is increased, thus also increasing its transverse velocity, the solution becomes more curved, but again the errors of the methods are about the same magnitude. Experiments with different grid sizes showed that as expected the global error contains leading terms of the order h^2 and k^2, except that the errors of the finite difference results are affected by a term of order h at the right hand boundary.

The scheme suggested above uses one-sided differences to approximate the boundary condition $y_s + y_t = 0$ in equation (3-6). A more accurate result is obtained by replacing (3-6) by a central difference formula

$$\frac{1}{8h}\left\{\delta_s y_{N,j+1} + 4\delta_s y_{Nj} + \delta_s y_{N,j-1}\right\} + \frac{1}{2k} \delta_t y_{Nj} = 0 . \quad (5\text{-}1)$$

This introduces extra unknowns $y_{N+1,j+1}$ outside the boundary, but these are easily eliminated between equation (5-1) and equation (3-5) with $i = N$. We then obtain a slightly different tridiagonal system of equations for the N unknowns $z_1, \ldots, z_N$.

This modification to the difference scheme was found to give improved accuracy for larger values of N, and experiments now showed the error decreasing by a factor of 4 when both h and k are halved.

When the time step was increased there was no sign of instability in either the finite difference or the characteristic method. This was expected; if we linearize the scheme (3-5), and replace the variable coefficients by constants of about the average size, the usual von Neumann analysis shows very easily that the simplified scheme is unconditionally stable for all values of h and k. This analysis is not conclusive, of course, but it is borne out by numerical experiments.

In terms of computer time, the characteristic method turns out to be slightly faster for small values of k, with k/h up to order unity, and slightly slower for k/h much greater than 1. The differences are not great, never more than about 50%, and neither program was carefully tuned. It is likely that more careful attention to the detail of the computation, and in particular to the various iterations involved, could affect the running times of the programs sufficiently to alter this comparison.

Some numerical results are given in Table 1 for illustration. The general conclusion is that for this problem the finite difference method and the method of characteristics give results of similar accuracy with similar computer time. The results are hardly affected by the very weak singularities. If the solution had singularities in the first derivatives, rather than the fourth derivatives, the conclusion would be very different, with the characteristic method having a clear advantage.

In the illustrative table the solution is nearly linear in s, with quite a small curvature y_{ss}. Other calculations, particularly with a higher frequency λ, have been found to give a more curved form to the cylinder. The numerical results from a series of calculations have provided useful information to the designers of the hydrophone system, and seem to agree well with observed behaviour.

Table 1

s	$y(s)$	Error in finite difference method $\times 10^4$	Error in characteristic method $\times 10^4$
0.00000	1.9900	0	0
0.03125	1.9815	0	1
0.06250	1.9724	-1	2
0.09375	1.9628	-1	2
0.12500	1.9527	-2	4
0.15625	1.9420	-3	3
0.18750	1.9310	-4	4
0.21875	1.9196	-6	5
0.25000	1.9080	-8	5

The solution $y(s)$ and the errors of the two numerical methods, at $t = 3$, using $N = 16$, $k = 0.02$; with the data $A = 0.35$, $s_1 = 0.25$, $\alpha = 1/30$, $\beta = 3$, $\lambda = 1$. The tabulated solution $y(s)$ was obtained with a finer grid, chosen to give an accuracy of at least four decimal places.

This problem was presented at one of the Oxford Study Groups with Industry. We are grateful to Plessey for permission to publish this paper.

REFERENCES

Ferriss, D.H. 1981, Numerical determination of the configuration of an underwater umbilical subjected to steady hydrodynamic loading, Parts 1 and 2. *National Physical Laboratory DNACS Report* 50/81.

Huffman, R.R. & Genin, J. 1971, The dynamical behaviour of a flexible cable in a uniform flow field. *Aeronautical Quarterly* **22**, 183-195.

12

A mathematical model of electromagnetic river gauging

P. JACOBY AND M. BAINES

1. INTRODUCTION

Electromagnetic river gauging makes use of the fact that a magnetic field will interact with a moving conductor of electricity passing through it, thus inducing an electromotive force. At a gauging station a coil situated underneath the river bed (or above a channel) gives rise to a magnetic field in the fluid. This field combines with the flow of the water to establish a potential difference across the river which can be measured via two probes, one on either bank. This chapter describes a mathematical model for use in the calibration of such a station. It is assumed that (1) the flow is incompressible, (2) the induced electric currents give rise to a negligible magnetic field, (3) the fluid obeys Ohm's Law and (4) the electrical conductivity is constant throughout the fluid. It can then be shown, Bevir [1961], that the gauging process is governed by the following equation which is known as Bevir's expression:

$$V = \int_{\Omega} \boldsymbol{v} \cdot (B \wedge \sigma \nabla u)\, d\Omega \qquad (1\text{-}1)$$

where V is the induced potential difference between the probes, $\boldsymbol{v}$ is the fluid velocity, B is the magnetic field, Ω is the entire volume of the fluid, σ is the conductivity of the water, and u is the potential which exists when a unit current flows

through the system under identical conditions to those prevailing
during flow measurement.

In order to evaluate Bevir's expression it is necessary to
carry out the following:

(1)   Calculation of the potential distribution. This obeys
        Laplace's equation which is solved numerically using
        the finite element method.

(2)   Normalization of the potential values to correspond
        with a unit current.

(3)   Evaluation of the magnetic field.

(4)   Incorporation of velocity data.

(5)   Evaluation of the expression (1-1) which has previously
        been rearranged into a more suitable form.

2.   THE FINITE ELEMENT METHOD

        The electric potential, u, obeys Laplace's equation

$$\nabla \cdot (\sigma \nabla u) = 0$$

where σ is the (possibly varying) electrical conductivity. If
boundary conditions are of Dirichlet or homogeneous Neumann type,
the solution to Laplace's equation, as is well known, can be obtained by minimization of the functional

$$I(u) = \tfrac{1}{2} \int_{\Omega} \sigma (\nabla u \cdot \nabla u) \, d\Omega \qquad (2\text{-}1)$$

where Ω is the entire conducting region. In the finite element
method nodal points are defined throughout the solution domain
and these are linked to form elements in space. The true solution u is then approximated by a piecewise function u^h given by

$$u^h = \sum_i u_i \phi_i \qquad (2\text{-}2)$$

where the $\{u_i\}$ are the nodal values of the finite element solution and the $\{\phi_i\}$ are local basis functions, usually low-degree
polynomials, each associated with a node and taking the value

unity at that node and zero at other nodes.

To minimize (2-1) we require that

$$\frac{\partial I}{\partial u_i} = 0 \qquad (i = 1, \ldots, n)$$

from which, using (2-1) and (2-2)

$$\sum_j \left\{ \int_\Omega \sigma (\nabla \phi_i \cdot \nabla \phi_j) d\Omega \right\} u_j = 0 \qquad (i = 1, \ldots, n)$$

or

$$Ku = 0 \qquad\qquad (2-3)$$

where K, known as the stiffness matrix, is given by

$$K_{ij} = \int_\Omega \sigma (\nabla \phi_i \cdot \nabla \phi_j) \, d\Omega = \sum_k \int_{\Omega_k} \sigma (\nabla \phi_i \cdot \nabla \phi_j) \, d\Omega_k, \qquad (2-4)$$

the $\{\Omega_k\}$ being those elements upon which both ϕ_i and ϕ_j are defined.

After imposing the Dirichlet boundary conditions, (2-3) becomes a soluble system of linear equations. Note that the solution produced by the finite element method satisfies the natural boundary condition

$$\frac{\partial u}{\partial n} = 0$$

in a weak sense over those portions of the domain boundary where no Dirichlet conditions are imposed (Strang & Fix [1973]). The finite element solution u^h can be shown to be the best fit in the energy norm to the true solution within the finite dimensional space represented by the basis functions.

3.   SOLUTION DOMAIN AND BOUNDARY CONDITIONS

In its basic form, the model consists of identical solid probes situated on either bank of a full river of trapezoidal cross-section surrounded by a conducting river bed. The water and the ground are assumed to have constant but different conductivities, and the air is assumed to be a perfect insulator. The 3-dimensional potential distribution with the probe at a fixed arbitrary potential is calculated by the finite element method.

The solution domain is shown in Figure 1.   Because of

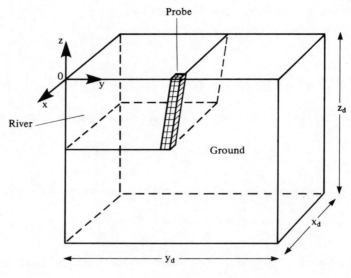

Figure 1

symmetry, only one quadrant of the region of interest is considered. A right-handed rectangular coordinate system is used
with the origin at the centre of the river in the plane of the
probe, the x-direction downstream, the y-direction towards the
left-hand bank and the z-direction vertically upwards.

The boundary conditions are as follows:

(1) $u = a$ on the probe. The arbitrarily chosen value of $a = 1$
     is used initially.

(2) $u(x, 0, z) = 0$. There is a probe at $u = -a$ on the oppo-
     site bank, and this condition follows from the resulting
     symmetry of the problem.

(3) $\left.\dfrac{\partial u}{\partial n}\right|_{z=0} = 0$, since air is assumed to be an insulator.

(4) $\left.\dfrac{\partial u}{\partial n}\right|_{x=0} = 0$, due to the symmetry of the problem.

In addition, a condition at the far boundary is required. If the
domain is sufficiently large for the potential to have decayed
to a negligible value on its outer boundaries, we can take

(5) $$u(x_d, y, z) = u(x, y_d, z) = u(x, y, z_d) = 0 \qquad (3\text{-}1)$$

as the extra condition. However, when the infinite boundary technique is used (see Section 6) this boundary condition can be omitted.

4.   THE MESH

The form of mesh which has been employed has as its principal feature a series of co-axial cylinders with bases or 'buckets' with the probe on the axis (see Figures 2 and 3).

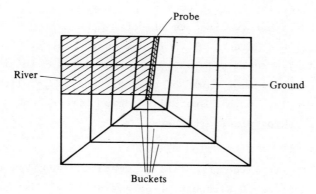

Figure 2

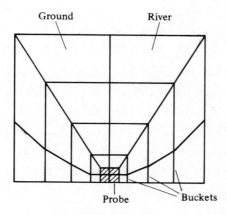

Figure 3

Nodal points are created at the intersection of these buckets
with sets of vertical and horizontal planes sub-dividing the
domain. The elements obtained are quadrilateral based prisms
with a node at each of the eight vertices. There are thus three
indices of mesh refinement — the number of vertical planes, denoted by IRX, the number of buckets, IRY, and the number of
horizontal planes, IRZ.

It was found that the finite element solution of this
problem is particularly sensitive to the number of buckets
employed, i.e. convergence occurs most rapidly with an increase
in IRY, the other two indices of refinement having by comparison a negligible effect. This is due to the fact that the true
solution decays rapidly in all directions in the vicinity of the
probe. The particular form of mesh was chosen to allow a high
density of nodes in this sensitive region without unnecessary
refinement taking place elsewhere.

5.   THE INFINITE SOLUTION METHOD

In reality, the potential u emanating from a source will
extend to infinity. Although u decays at large distances, the
boundary condition $u = 0$ imposed on the outer boundaries of a
finite solution domain is therefore not strictly correct. A
method of simulating the infinite boundary condition on finite
boundaries has been incorporated into the program so that the
unlimited extent of the potential field can be modelled.

This section summarizes a paper by Silvester *et al.* [1977]
who devised the technique for Laplace's equation in 2 dimensions
where, as here, the solution extends over an infinite region but
is of interest in only a finite domain surrounding the source.

An annular region, Ω_1, is formed around the solution domain R (Figure 4) in such a way that every point (x_0, y_0) on the
boundary r_0 of R can be mapped on to a point (x_1, y_1) on the
outer boundary Γ_1 of Ω_1 by the relationship

$$(x_1, y_1) = k(x_0, y_0) \qquad (5\text{-}1)$$

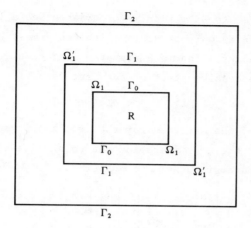

Figure 4

where k is a constant, known as the scale factor. In particular, the n nodes on the boundary of R are mapped on to nodes on Γ_1. The two sets of nodes are used as the vertices of a convenient set of elements subdividing Ω_1 and the $2n \times 2n$ finite element matrix, S^1, can then be assembled for Ω_1. If Γ_1 is mapped on to a third, more distant boundary, Γ_2, by a similar transformation to (5-1) and the new annular region, Ω_1', so formed is similarly divided into elements, then all element dimensions in Ω_1' will by k times as large as their equivalents in Ω_1. This being the case, it can be shown that the matrix $S^{1\prime}$ for Ω_1' is identical to S^1. Combination of S^1 and $S^{1\prime}$ and elimination of common nodes will give rise to a new $2n \times 2n$ matrix, S^2, describing the region Ω_2 which is the union of Ω_1 and Ω_1'. S^2 contains only the linkages of nodes on the domain boundary, Γ_0, and the outer boundary, Γ_2. The relationship between points on Γ_0 and Γ_2 is given by

$$(x_2 , y_2) = k^2 (x_0 , y_0) . \qquad (5-2)$$

If a new region, Ω_2', is then created whose dimensions are k^2 times as large as those of Ω_2, then the matrix $S^{2\prime}$ is identical to S^2 and their combination, as before, describes a still larger composite region. This process may continue, each iteration giving rise to a $2n \times 2n$ matrix which describes a larger exterior layer in terms only of nodes on the outermost boundary and the domain boundary. The main diameter of the outer layer is increased at successive steps by the factors

$$k , k^2 , k^4 , k^8 , k^{16} , \ldots \qquad (5\text{-}3)$$

A combination of the matrices for the two annuli can be performed as follows. At the jth iteration the nodal equations for Ω_j (see Figure 5) are given by $S^j \phi = 0$ or

$$\begin{bmatrix} S^j_{11} & S^j_{12} \\ S^j_{21} & S^j_{22} \end{bmatrix} \begin{bmatrix} \phi_1 \\ \phi_2 \end{bmatrix} = 0$$

where S^j_{11} etc. are $n \times n$ submatrices, ϕ_1 are the nodes on Γ_0 and ϕ_2 are the nodes on Γ_j.

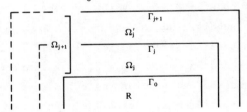

Figure 5

Since S^j is identical to $S^{j'}$, the equations for Ω_{j+1} are

$$\begin{bmatrix} S^j_{11} & S^j_{12} & 0 \\ S^j_{21} & S^j_{22} + S^j_{11} & S^j_{12} \\ 0 & S^j_{21} & S^j_{22} \end{bmatrix} \begin{bmatrix} \phi_1 \\ \phi_2 \\ \phi_3 \end{bmatrix} = 0 , \qquad (5\text{-}4)$$

where ϕ_3 are the nodes on Γ_{j+1}. From this we obtain

$$\phi_2 = -A^j S^j_{21} \phi_1 - A^j S^j_{12} \phi_3 \qquad (5\text{-}5)$$

where

$$A^j = (S^j_{22} + S^j_{11})^{-1}$$

and substituting for ϕ_2 in (5-4) gives

$$(S_{11}^j - S_{12}^j A^j S_{21}^j)\phi_1 - S_{12}^j A^j S_{12}^j \phi_3 = 0$$

$$-S_{21}^j A^j S_{21}^j \phi_1 + (S_{22}^j - S_{21}^j A^j S_{12}^j)\phi_3 = 0 .$$

Having thus eliminated the nodes on Γ_j, the reduced matrix for Ω_{j+1} is given by

$$S^{j+1} = \begin{bmatrix} S_{11}^j & 0 \\ 0 & S_{22}^j \end{bmatrix} - \begin{bmatrix} S_{12}^j A^j S_{21}^j & S_{12}^j A^j S_{12}^j \\ S_{21}^j A^j S_{21}^j & S_{21}^j A^j S_{12}^j \end{bmatrix} . \qquad (5-6)$$

This recurrence relationship means that only the matrix for the first exterior layer need be assembled in the normal manner. All succeeding matrices are generated from elements of the previous one by (5-6) and the layers themselves need not be explicitly constructed. After the nth and final iteration has been per-formed, the submatrix S_{11}^n, describing the exterior linkages of nodes on the domain boundary, is used to augment the correspond-ing entries of the domain stiffness matrix. The final solution within the domain thus takes account of its behaviour over a large region outside.

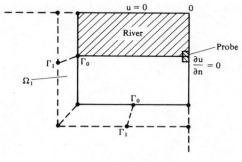

Figure 6

It turns out that Silvester's method is well suited for use in the present problem. A layer (Figure 6), partially sur-rounding the solution domain, is created on the three faces of the domain where the approximate boundary condition $u = 0$ (see Section 3) had previously been applied. A three-dimensional mapping must here be used whereby each node, (x_0, y_0, z_0), on the relevant part, Γ_0, of the domain boundary is mapped on to the enclosing surface, Γ_1, by

$$(x_1, y_1, z_1) = k(x_0, y_0, z_0) \qquad (5\text{-}7)$$

thus forming the first boundary layer, Ω_1. Elements of similar shape to those used in the main domain are generated over Ω_1, values of conductivity are assigned to them, and the matrix S^1 is assembled in the normal fashion. In order to generate matrices for succeeding layers, the recurrence relation (5-6) must be modified. In 3 dimensions, the stiffness matrix entries for the Laplacian take the form

$$\left\{K_{ij}\right\}^{\Omega_k} = \iiint\limits_{\Omega_k} \left(\frac{\partial \phi_i}{\partial x} \frac{\partial \phi_j}{\partial x} + \frac{\partial \phi_i}{\partial y} \frac{\partial \phi_j}{\partial y} + \frac{\partial \phi_i}{\partial z} \frac{\partial \phi_j}{\partial z} \right) dx\, dy\, dz \quad (5\text{-}8)$$

and from inspection of this expression it can be seen that, if an element Ω_l has all dimensions C times as large as in Ω_k, where C is a constant, then

$$\left\{K_{ij}\right\}^{\Omega_l} = C\left\{K_{ij}\right\}^{\Omega_k}. \qquad (5\text{-}9)$$

For the method in 3 dimensions, (5-4) then becomes

$$\begin{bmatrix} S_{11}^j & S_{12}^j & 0 \\ S_{21}^j & S_{22}^j + K_j\, S_{11}^j & K_j\, S_{12}^j \\ 0 & K_j\, S_{21}^j & K_j\, S_{22}^j \end{bmatrix} \begin{bmatrix} \phi_1 \\ \phi_2 \\ \phi_3 \end{bmatrix} = 0 \qquad (5.10)$$

where

$$K_j = k^{2^{j-1}} \qquad (5\text{-}11)$$

is the scale factor at the jth iteration of the scheme. The recurrence relationship for the matrices is now given by

$$S^{j+1} = \begin{bmatrix} S^j_{11} & 0 \\ 0 & K_j S^j_{22} \end{bmatrix} - \begin{bmatrix} S^j_{12} B^j S^j_{21} & K_j S^j_{12} B^j S^j_{12} \\ K_j S^j_{21} B^j S^j_{21} & K^2_j S^j_{21} B^j S^j_{12} \end{bmatrix} \qquad (5\text{-}12)$$

where

$$K_j = K^2_{j-1}$$

and

$$B^j = (S^j_{22} + K_j S^j_{11})^{-1}$$

A sequence of matrices can thus be generated for an expanding 3 dimensional exterior layer and the matrix from the final iteration used, as before, to amend the solution within the domain.

The method was designed to deal with a uniform medium and a disadvantage of its application to this problem is that, as Figure 6 shows, it is impossible to accurately model the river in the exterior layer by having element boundaries coincide with its banks. As the layers move outwards, all element dimensions must increase, with the result that, at most, only part of any element will consist of the river. In fact expansion is very rapid:- substituting a value of $k = 1.5$ as the initial scale factor in (5-3) (this is a value recommended by Silvester *et al.* [1977]) the outer diameter of the layer and the linear size of any element will be increased at successive steps by the factors

1.5, 2.25, 5.06, 25.6, 65.7, 43.1 $\times$ 10^4, ...

Therefore, even an element which in the first layer consisted wholly of the river will, after a few iterations, be extended into one occupied predominantly by the ground. Since only one value of conductivity can be assigned to an element and the assignment is performed once only when constructing the first layer, we have allotted the bed conductivity value to all exterior elements.

Results of an application of the method are summarized in Table 1. The domain used in obtaining these results extended two river's width downstream of the probe and between one and

Table 1. Effect of boundary iterations on the solution inside the domain. N is the number of iterations, u is the solution value and P_1 and P_2 are distinct points in the river, P_2 being on the domain boundary.

N	$u(P_1)$	$u(P_2)$
0	0.2320	0.0000
1	0.2413	0.0340
2	0.2430	0.0380
3	0.2437	0.0398
4	0.2439	0.0402
5	0.2439	0.0403

two river's widths into the ground beside and underneath the water. A value of 1.5 for the initial scale factor was used. It can be seen that the nodal solution values inside the domain converge to about four figures after five iterations, the converged values representing an increase of about 5% on the original solution which was forced to zero on the domain boundary.

6.   UNIT CURRENT AS A BOUNDARY CONDITION

Sections 2 and 3 describe the solution of the problem when a unit potential is imposed upon the probe. However, Bevir's expression (1-1) requires the solution when a unit current flows through the system. One way of specifying unit current is to solve the problem for a unit potential on the probe and then adjust the potential values so that they correspond to a unit current issuing from the probe. The exact current, I, is given by

$$I = \int_{\Gamma} \sigma \frac{\partial u}{\partial n} \, d\Gamma , \qquad (6-1)$$

where Γ is a surface enclosing the probe and n is the outward normal direction to Γ. Since the potential at any point is

proportional to the total current flow through the system, normalization to unit current is performed by simply dividing all
potential values by the measured current. For the finite element
solution the integral in (6-1) is replaced by an approximating
sum but the linear relationship still holds.

A question remains as to the choice of the measuring surface. Gauss's theorem states that

$$\int_{\Gamma} \boldsymbol{J} \cdot \boldsymbol{n} \; d\Gamma = \int_{\Omega} \boldsymbol{\nabla} \cdot \boldsymbol{J} \; d\Omega, \tag{6-2}$$

where $\boldsymbol{J}$ is the current density, Γ a closed surface and Ω the
region bounded by Γ. Since $\boldsymbol{J} = \sigma \nabla u$ it follows that

$$\int_{\Gamma_1} \sigma \frac{\partial u}{\partial n} \; d\Gamma_1 - \int_{\Gamma_2} \sigma \frac{\partial u}{\partial n} \; d\Gamma_2 = \int_{\Omega} \boldsymbol{\nabla} \cdot (\sigma \nabla u) \; d\Omega \tag{6-3}$$

for any two surfaces, Γ_1 and Γ_2, enclosing the probe. If u
satisfies Laplace's equation the currents measured across Γ_1 and
Γ_2 will thus be equal and for a fully converged numerical solution of Laplace's equation the choice of current measuring
surface will be immaterial. However, this will not be the case
when the solution is still an approximation. The currents measured on three different surfaces (coinciding with element
boundaries) are substantially different even for a refined mesh,
and if they do converge to the same value, progress is extremely
slow. However, finite element theory indicates that this discrepancy does not necessarily represent a serious inaccuracy in
the solution itself at the last stage of refinement. In this type
of problem, with linear basis functions, the convergence of the
finite element solution to the true solution is $O(h^2)$ where h is
the element side, (Strang & Fix [1973]). The convergence of the
first derivative is however only $O(h)$. Although these results
are strictly only applicable to regular meshes, the principle of
slower derivative convergence will still apply. It implies that
the measured current, which involves calculation of first

derivatives of the finite element solution, is likely, at a given stage of mesh refinement, to be considerably less accurate than values of the solution itself. However, the theory also predicts that at certain points within the domain, called superconvergence points, the derivatives have a higher order of convergence than elsewhere (Zienkiewicz [1977], Lesaint & Zlamal [1979]); in particular, the accuracy of the original solution is recovered for the first derivative at the $1 \times 1 \times 1$ Gauss quadrature points which are the centroids of the elements. This suggests that in order to obtain greater accuracy for the measured current the measuring surface should be chosen in such a way that it links these superconvergence points. Calculation of the current by (6-1) can then be performed by a quadrature rule which requires evaluation of the derivative at these points only.

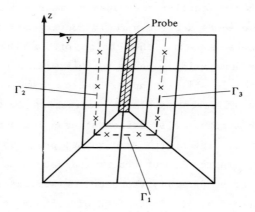

Figure 7

Implementation of this suggested procedure is achieved as follows. The surface Γ_1 has $z = \text{constant}$ throughout. Therefore, within elements Ω^E containing Γ_1 (see Figure 7) the contribution $I\phi_i$ from the basis function ϕ_i to the current is given by

$$I\phi_i = \int_{\Gamma_1^E} \sigma \, \frac{\partial \phi_i}{\partial n} \, d\Gamma_1^E = -\int \int_{\Gamma_1^E} \sigma \, \frac{\partial \phi_i}{\partial z} \, dx \, dy \,. \qquad (6\text{-}4)$$

(The sub-surfaces Γ_2 and Γ_3 require a similar but slightly special treatment.) The integrals are evaluated using 1×1 Gauss quadrature, for which the integration point is the centroid of the element.

The calculation takes each element which contains the current measuring surface and calculates the contribution, $I\phi_i$, to the current from each basis function of the element. Then, after solution of the potential distribution they are amalgamated to give a value of the current flow, I, using

$$I = \sum_i u_i \, I\phi_i \,, \qquad (6\text{-}5)$$

where u_i is the value of the solution at the node associated with ϕ_i. All solution values are then divided by I giving the distribution which corresponds to a unit current.

This more sophisticated technique of current measurement was tested by performing the measurement across a number of surfaces in a test domain, each surface consisting of a linkage of superconvergence points. The behaviour of the current with mesh refinement was taken for four different such surfaces. After the final stage of refinement the values have converged to very nearly the same point, there being a discrepancy of less than 1% between any two of them. This is in contrast to the unreliable results which were obtained when the measuring surface was chosen to consist of element boundaries. The improved measuring technique has consequently been included in the method. The surface used is that which links the centroids of the first layer of elements surrounding the probe.

7.   VARIATIONS TO THE BASIC MODEL

As well as the standard situation described above, it is desirable to be able to model a variety of physical situations

likely to be present at a gauging station, either separately or
in combination.

(a)   *The Half-Full River*

     A completely full river is unusual and it is therefore
important that the program be able to model the more likely situation of a partly-full river. The technique used is to reduce
the size of the solution domain so that its boundary coincides
with that of the water surface. Figure 8(a) shows the $x = 0$ plane
of such a reduced domain. Mesh nodes in the water are generated
only at and below the water level. No extra boundary conditions
are imposed, so the natural condition at the edge of the domain,
$\partial u / \partial n = 0$, prevails, corresponding to the correct situation of
no current flow from the water into the air.

(b)   *Insulation Region*

     An insulating region placed directly behind the probes
will significantly increase the output from the gauging station.
In order to simulate this situation a portion of the solution
domain, corresponding to the insulating region, is removed. The
desired natural boundary condition of zero current flow automatically applies across the river/insulator and ground/insulator
interfaces. An example of this is shown in Figures 8b, c.

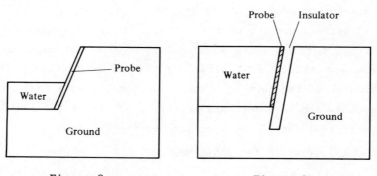

Figure 8a   Figure 8b

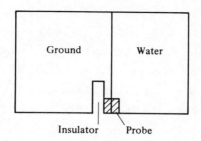

Figure 8c

(c)   *Irregular River Bed*

Where gauging stations are installed on natural channels, an uneven river bed is likely to be present and we incorporate this feature as follows. Suppose initial data is supplied giving the depth of the bed at the point where it joins the bank. The mesh generating routine calculates the y co-ordinates of the nodes which constitute the river bed, and outputs the positions at which bed depth values are required. Upon the input of this data, the routine is able to assign the correct co-ordinates to these nodes and thus model the actual bed profile. Figure 9 shows how an irregular bed might be approximated. This facility highlights the advantage of using the finite element method in problems such as this where irregular shapes are a feature of the domain.

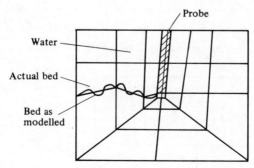

Figure 9

(d)   *The Probe*

The solid probe with three finite dimensions is set into the river bank (i.e. with half of its cross-section inside the bank and half in the water) and extending downwards to the bed. To coincide with the shape of the elements, the probe must have a rectangular cross-section, and therefore an actual probe with say, circular cross-section would only be approximated by the program. The approximation can be made in a variety of ways: for instance the side of the square can be set equal to the diameter of the actual probe, or their cross-sectional areas can be matched, or their circumferences set equal to one another.

A modification to the model allows it to cater for a probe buried wholly inside the river bank and extending downwards parallel to the edge of the bank. Data specifying the probe's position is input, and if necessary element boundaries are adjusted so that they coincide with both the river bank and the outline of the probe in its new position. Element conductivities are assigned according to whether the element lies in the water or in the ground.

(e)   *Variable Bed Conductivity*

The assumption of a constant river bed conductivity, σ_G, is unlikely to hold in practice. Layers bordering the water will probably be wetted and therefore have a higher conductivity than regions further away from the river and this will affect the potential distribution. A suggested adaptation of the model to this end consists of amending the routine which assigns values of

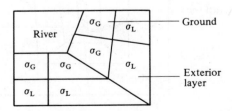

Figure 10

σ to the elements so that, assuming the infinite boundary technique is being used, (Section 5), a third, lower value of σ is allotted to those elements generated in the exterior layer (Figure 10).

8.   REARRANGEMENT OF BEVIR'S EXPRESSION

Bevir's expression (1-1) for the induced e.m.f.

$$V = \int_{\Omega} \boldsymbol{v} \cdot (\boldsymbol{B} \wedge \sigma \nabla u) \, d\Omega \ , \tag{8-1}$$

can be arranged to give

$$V = \int_{\Omega} \sigma \nabla u \cdot (\boldsymbol{v} \wedge \boldsymbol{B}) \, d\Omega \ . \tag{8-2}$$

Applying Green's theorem for volume integrals we then obtain

$$V = \int_{\Omega} \sigma u (-\operatorname{div}(\boldsymbol{v} \wedge \boldsymbol{B})) \, d\Omega + \int_{\Gamma} \sigma u \, (\boldsymbol{v} \wedge \boldsymbol{B}) \cdot \boldsymbol{n} \, d\Gamma$$

$$= \int_{\Omega} \sigma u \, (\boldsymbol{v} \cdot \operatorname{curl} \boldsymbol{B} - \boldsymbol{B} \cdot \operatorname{curl} \boldsymbol{v}) \, d\Omega + \int_{\Gamma} \sigma u (\boldsymbol{v} \wedge \boldsymbol{B}) \cdot \boldsymbol{n} \, d\Gamma \tag{8-3}$$

where Γ is the surface of Ω.

A steady linear current, such as we have in the coil, gives rise to an irrotational magnetic field, so that

$$\operatorname{curl} \boldsymbol{B} = 0$$

and (1-1) reduces to

$$V = \int_{\Omega} -\sigma \, u \, (\boldsymbol{B} \cdot \operatorname{curl} \boldsymbol{v}) \, d\Omega + \int_{\Gamma} \sigma \, u (\boldsymbol{v} \wedge \boldsymbol{B}) \cdot \boldsymbol{n} \, d\Gamma \ . \tag{8-4}$$

If we now make the assumption that the water velocity, $\boldsymbol{v}$, possesses only a component in the x (downstream) direction, then, denoting v_x by v, (8-4) becomes

$$V = \int_{\Omega} \sigma \, u \left(B_z \frac{\partial v}{\partial y} - B_y \frac{\partial v}{\partial z} \right) d\Omega + \int_{\Gamma_1} \sigma \, u v B_y \, d\Gamma_1$$

$$- \int_{\Gamma_2} \sigma \, u v B_z \, d\Gamma_2 - \int_{\Gamma_3} \sigma \, u v B_y \, d\Gamma_3 \tag{8-5}$$

where Γ_1 is the river surface, Γ_2 the bank (here assumed to be

perpendicular to the river surface) and Γ_3 the bed. This is the
reformulated Bevir's expression that is used to calculate the
output of a station.

If the water velocity, v, is uniform the first term of
(8-5) disappears. In this case only surface integrals need be
evaluated in the final estimation of the e.m.f. Note that this
assumption is only consistent with non-viscous flow. Otherwise
the boundary conditions for viscous flow must be used, namely,
that the velocity is zero where the fluid comes into contact with
the stationary bed. In this case the third and fourth terms of
(8-5) disappear but the derivatives of v are now non-zero and
the first term of (8-5) must be considered.

We have already noted that the derivatives of a finite
element solution have a lower order of accuracy than the solution itself, except at certain points of superconvergence. However, a result of the above rearrangement is that derivatives of
the potential, u, have been removed from the expression. In
evaluating the volume integral of (8-5) by numerical integration,
values of the variables in the integrand can thus be sampled at
any point in the domain without loss of confidence in their
accuracy. This has the advantage that a higher order of quadrature can be used and variations in the velocity and magnetic
field can be more accurately modelled.

The first term of (8-5) is evaluated by piecewise $2 \times 2 \times 2$
Gauss quadrature over the same elements in the river as were
used in solving the potential problem. The surface integral terms
are similarly evaluated using 2×2 quadrature. As well as the
potential, values of the magnetic field components and the
velocity or its derivatives are required at the integration
points. A known formula is available for each component of the
magnetic field produced by a rectangular coil. However, coils
installed at existing stations are in the shape of a trough with
'arms' extending upwards on either side of the river (see Figure

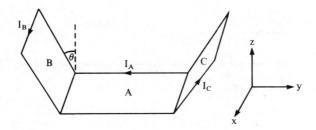

Figure 11

11).   In order to derive the field for a coil of this shape, it
may be considered as consisting of three separate units, A, B
and C, each of which is rectangular. Then, by suitable orientation of an equal current in each unit, the field produced by the
composite coil will be given by the sum of the fields from the
three units, each of which may be calculated from the available
formula by a simple transformation of axes and coordinates.

9.   TEST PROBLEMS

(a)   The model can be checked by investigating a test problem
with an available analytic solution for the potential.   The main
purpose of this operation is to examine the rate of convergence
of the finite element solution to the true solution for various
mesh refinement strategies.   The test problem consists of finding the potential distribution in a fully insulated channel with
probes of finite cross-section on either bank.   The probes are at
a potential of equal magnitude but opposite sign.

        Because of the insulation of the channel, the potential
has no vertical dependence and the resulting two dimensional
problem can be solved using the method of images.   Consider two
point sources of strengths m and $-m$ on either side of a two
dimensional channel (see Figure 12).   The insulation imposes a
symmetry on the solution within the channel such that it is
identical with that produced by an infinite number of sources
with alternating signs spaced equally along a line.   With the

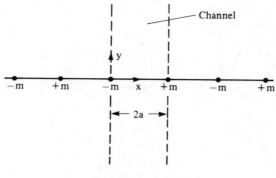

Figure 12

origin at the centre of the channel, it can be shown, using a similar problem in hydrodynamics (Milne-Thompson [1968]) that the complex potential distribution is given by

$$w = -m \, \ell n \left(\sin \frac{\pi}{4a} \, (z-a) \right) + m \, \ell n \left(\sin \frac{\pi}{4a} \, (z+a) \right) \qquad (9\text{-}1)$$

where $z = x + iy$.

The electric potential, u, can then be obtained, namely,

$$u = \text{Re} \, (w) = m \, \ell n \left| \frac{\sin \frac{\pi}{4a} \, (z+a)}{\sin \frac{\pi}{4a} \, (z-a)} \right| = \frac{m}{2} \, \ell n \left(\frac{\sin^2 \frac{\pi}{4a} (x+a) + \sinh^2 \frac{\pi y}{4a}}{\sin^2 \frac{\pi}{4a} (x-a) + \sinh^2 \frac{\pi y}{4a}} \right)$$

$$(9\text{-}2)$$

Of course, the potential is infinite at the point source. In order for (9-2) to describe the distribution produced by probes at unit potential, we specify a region surrounding the source to be occupied by the probe. Then taking a point (x_0, y_0) on the surface of the probe and setting $u = 1$, $x = x_0$ and $y = y_0$, (9-2) can be solved for m. This value of m is now substituted back into (9-2).

For (9-2) to be the exact solution of a problem modelled here the surface of the probe should coincide with an equipotential defined by (9-2). However, a constraint of the mesh is that any surface it describes must be rectangular, and this is clearly not a property of the equipotentials. To overcome this, the values of the potential imposed at different points on the rectangular probe were allowed to vary slightly in accordance with (9-2). The numerical solution so obtained should then be directly comparable, throughout the channel, with the analytic expression.

The analytic solution (9-2) exhibits a singularity at the point $(a,0)$ which is the centre of the probe. The solution region for this problem is entirely exterior to the probe; however, if the probe is small, the area in the vicinity of its surface is sufficiently close to the singularity for the solution to vary rapidly there. Special treatment is required in the numerical modelling of this type of solution, and there are two basic approaches which can be used in the finite element method (Whiteman & Akin [1978]). One is to augment the basis functions so that they approximate the (known) form of the singularity. The other approach, and the one followed here, is to adopt some strategy of local mesh refinement where elements in the neighbourhood of the singularity are made sufficiently small for the basis functions to model the rapidly varying solution in this region.

A feature of the type of mesh used (see Section 4) is that the crucial aspects of refinement are the number and spacing of buckets surrounding the probe. A refinement strategy therefore consists of the positioning of additional buckets in accordance with a particular spacing function of form

$$h_n = f(n) \qquad\qquad (9-3)$$

where h_n is the width of the nth bucket away from the probe (Figure 13). The convergence of the numerical solution to the

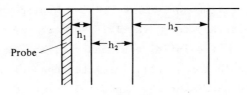

Figure 13

true solution was investigated for a number of different refinement strategies all of which exhibited the property of increasing
element width in the direction away from the probe.

The strategy which led to most rapid convergence was
'quadratic' spacing given by

$$h_n = \text{const} \times n^2 . \tag{9-4}$$

The convergence performance of the strategy is summarized in
Table 2.

Table 2. Convergence with quadratic spacing. n is
the number of buckets, E_n is the L_2 error norm
given by $E_n = \left(\frac{1}{V} \int_\Omega |u - u^h|^2 \, d\Omega\right)^{\frac{1}{2}}$ where V is the
volume of the solution region, Ω. The column E_n'
gives values of a faster converging sequence generated
by applying Aitken's Δ^2 process to values of E_n.

n	E_n	E_n'
3	0.044337	
4	0.027908	0.014420
5	0.020501	0.011063
6	0.016351	0.007893
7	0.013567	0.005616
8	0.011505	

The L_2 norm

$$\|u\| = \left(\frac{1}{V} \int_\Omega |u|^2 \, d\Omega\right)^{\frac{1}{2}} \tag{9-5}$$

of the exact solution (9-2) was calculated and is given by
$\|u\| = 0.113137$. The error in the numerical solution after the
final quadratic refinement (Table 2) is thus approximately 10%.
This level of accuracy for a fairly fine mesh may appear disappointing. However, the infinite boundary technique (see Section
5) is not suitable for a fully insulated channel and was therefore not used in obtaining the numerical solution. Instead, the

potential was forced to zero at a finite distance downstream of the probe whereas the exact solution is of infinite extent in the channel. A solution which incorporates this infinite boundary technique can be expected to be more accurate for an equivalent level of refinement.

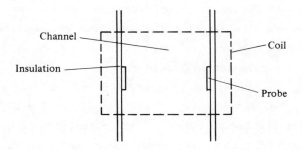

Figure 14

A working physical model of an electromagnetic gauging station has been installed at the Hydraulics Research Station [1979]. It is therefore possible to assess the performance of the model by comparing the e.m.f. output predicted by it with actual data obtained from the working model. The Wallingford station consists of an insulated channel of rectangular cross-section with plate probes on either side (see Figure 14). A square coil is in position above the channel. Relevant dimensions are as follows: — width of channel — 1.475 m; depth of channel — 0.737 m; width of probe — 0.125 m; extent of insulation — ~4.5 m; diameter of coil — 1.854 m; distance of coil above top of channel — 0.102 m. The station output, V, is calculated by evaluating the following expression (see Section 8) involving surface integrals.

$$V = \int_{\Gamma_1} \sigma \, uv B_y \, d\Gamma_1 - \int_{\Gamma_2} \sigma \, uv B_2 \, d\Gamma_2 - \int_{\Gamma_3} \sigma \, uv B_y \, d\Gamma_3 . \quad (9-6)$$

Γ_1 is the water surface, Γ_2 the banks, and Γ_3 the bed of the channel. The water velocity, v, is assumed constant. Five test runs at the Wallingford station were simulated by the program, and results are shown in Table 3. A fairly fine mesh was used with $IRX = 1$, $IRY = 8$ and $IRZ = 2$ (see Section 4). Considering that the model assumed a constant water velocity and total insulation of the channel, the level of accuracy of these results is highly encouraging. Furthermore, the accuracy is fairly consistent over the range of operation of the station.

A repeat simulation of the same test runs was performed, this time using the original volume integral form (1-1) of Bevir's expression, to calculate the output.

Table 3. Comparison of predicted and observed output. h is the water depth, I the current in the coil, v the mean water velocity, V the observed voltage output and V_p the output predicted by the program.

Run	h(mm)	I(A)	v(m/s)	V(μV)	V_p(μV)	Error (%)
1	169.9	1159	0.025	10.68	10.14	− 5.1
2	204.5	1151	0.094	41.11	38.95	− 5.3
3	360.6	1153	0.128	64.60	60.60	− 6.2
4	643.8	1164	0.237	136.23	147.20	+ 8.1
5	364.5	1162	0.422	210.34	202.10	− 3.9

Table 4. Comparison of predicted and observed output with alternative form of calculation. V is the observed voltage output and V_p that predicted by the program.

Run	V(μV)	V_p(μV)	Error (%)
1	10.68	10.06	− 5.8
2	41.11	38.64	− 6.0
3	64.60	60.06	− 7.0
4	136.23	145.38	+ 6.3
5	210.34	200.20	− 4.8

The same mesh was employed for these repeat runs. Results are shown in Table 4. It can be seen that for four out of the five runs the rearranged form of Bevir's expression (Table 3) gives more accurate results, but the difference is very small. However, it is envisaged that the advantage of the rearrangement of the expression will be manifested more clearly when a fairly turbulent flow in a natural channel is simulated and variations in water velocity can be modelled more accurately.

10.   CONCLUSION

The aim of this work was to provide an accurate means of calibration of an electromagnetic river gauging station. Although only a few of these exist, it is also applicable to flows in ducts and channel flows in sewerage stations. The primary purpose and use of the study awaits the construction and use of further gauging stations on rivers.

ACKNOWLEDGEMENTS

This work was carried out while one of us (Peter Jacoby) was working at the University of Reading under contract from the Water Research Centre, Medmenham. Thanks are due to the WRC for permission to publish this work and in particular to D. Oakes of the WRC for his advice and help.

REFERENCES

Bevir, M. 1961, Induced voltage electromagnetic flowmeters. Ph.D. thesis, University of Warwick.

Hydraulics Research Station, Wallingford. 1979, Calibration of electromagnetic gauging station. Report No. DE 48.

Lesaint, P. & Zlamal, M. 1979, Superconvergence of the gradient of finite element solutions. *RAIRO Analyse Numerique*, **13**, 139-166.

Milne-Thomson, L. 1968, *Theoretical Hydrodynamics*. 5th edition. Macmillan.

Silvester, P., Lowther, D., Carpenter, C. & Wyatt, E. 1977, Exterior finite elements for two-dimensional field problems with open boundaries. *Proc. IEE*, **124**, 1267-1270.

Strang, G. & Fix, G. 1973, *An Analysis of the Finite Element Method.* Prentice-Hall.

Whiteman, J. & Akin, J. 1978, Finite elements, singularities and fracture. Brunel University, Institute of Computational Mathematics Internal Report.

Zienkiewicz, O. 1977, *The Finite Element Method.* 3rd edition. McGraw-Hill.